MAKING YOUR N GAUGE RAILWAY MORE REALISTIC

For Mary

MAKING YOUR N GAUGE RAILWAY MORE REALISTIC

In the BR Blue and Sectorisation Eras

Richard Middleton

PEN & SWORD
TRANSPORT

AN IMPRINT OF PEN & SWORD BOOKS LTD.
YORKSHIRE – PHILADELPHIA

First published in Great Britain in 2024 by
Pen and Sword Transport
An imprint of
Pen & Sword Books Ltd.
Yorkshire – Philadelphia

Copyright © Richard Middleton, 2024

ISBN 978 1 39906 724 9

The right of Richard Middleton to be identified as author of this work has been asserted by him in accordance with the Copyright, Designs and Patents Act 1988.

A CIP catalogue record for this book is available from the British Library.

All rights reserved. No part of this book may be reproduced or transmitted in any form or by any means, electronic or mechanical including photocopying, recording or by any information storage and retrieval system, without permission from the Publisher in writing.

Typeset in 11.5/14 Palatino
by SJmagic DESIGN SERVICES, India.

Printed and bound by Printworks Global Ltd, London/Hong Kong.

Pen & Sword Books Ltd. incorporates the imprints of Pen & Sword Books: After the Battle, Archaeology, Atlas, Aviation, Battleground, Discovery, Family History, History, Maritime, Military, Naval, Politics, Railways, Select, Transport, True Crime, Fiction, Frontline Books, Leo Cooper, Praetorian Press, Seaforth Publishing, Wharncliffe and White Owl.

For a complete list of Pen & Sword titles please contact

PEN & SWORD BOOKS LIMITED
George House, Units 12 & 13, Beevor Street, Off Pontefract Road,
Barnsley, South Yorkshire, S71 1HN, England
E-mail: enquiries@pen-and-sword.co.uk
Website: www.pen-and-sword.co.uk

or

PEN AND SWORD BOOKS
1950 Lawrence Rd, Havertown, PA 19083, USA
E-mail: uspen-and-sword@casematepublishers.com
Website: www.penandswordbooks.com

Contents

Acknowledgements 9
Preface 10
The Scope of the Book 10

Chapter 1	The Railway As a System – Purpose, Destination, Traffic, Goods, Region and Geography 11	
	The BR Blue and Sectorisation Eras 12	
	Setting the Scene 12	
	Reality versus Fantasy 12	
Chapter 2	Train Formations 14	
	Total Operations Processing System (TOPS) and Vehicle CARKND 14	
	How to Understand Freight Vehicle TOPS Codes 15	
	Vehicle Compatibility – Couplers, Jumpers, Barrier, and Translator Vehicles 16	
	Train Marker Lighting, Lamps and Headcodes 20	
	Train and Brake Van Lighting 21	
	Selecting Appropriate Traction for Your Train 23	
	Other Compatibility Considerations – ETH Index and Train Brake Force 24	
	A Word on Uniformity 26	
	Passenger Train Formations 27	
	Intercity and Cross-Country workings 27	
	High Speed Train 28	
	Secondary Workings 30	
	Special Passenger Trains 32	
	Freight Train Formations 35	
	The Correct Use of Brakevans 36	
	Coal 36	
	Steel 41	
	Petroleum 42	
	Cement 46	

	Aggregates and Other Mineral Traffic	48
	Grain	51
	Chemicals	51
	Automotive	53
	Commercial Explosives	56
	The Air Braked Network (ABN) and Speedlink	57
	Parcels and Newspapers	58
	Mail and The Travelling Post Office (TPO)	60
	Departmental and Service Trains	62
	Intermodal	65
	Other Traffic	66
	Snowploughs	67
	Nuclear Trains	69
	Breakdown Trains	72
	Test Trains	74
Chapter 3	Loads and the Loading of Wagons	77
	General Rules	77
	Pipes/Tubes	78
	Ingots and Other Concentrated Loads	80
	Steel – Coil	83
	Steel – Sheet and Plate	85
	Steel – Rails	85
	Scrap Metal	89
	Timber	89
	Bogies and Wheelsets	91
	Reeled Paper	94
	Road Vehicles and Plant	95
	Pre-Assembled Track Panels	97
	Sleepers	98
	Cable Drums	99
	Spoil and Ballast	101
	Loose Materials	102
	Coal	103
	Special Loads	104
Chapter 4	Civil Engineering and the Permanent Way	106
	The Permanent Way	106
	Basic Track Layouts and Nomenclature	107
	Rail	109
	Sleepers and Ballast	110
	Switches and Crossings	111
	Catch and Trap Points	114
	Point Motors, Levers and Groundframes	115

	Expansion Joints	116
	Other Permanent Way Features	116
	On the Model – Operation and Appearance	118
	Railway Electrification	120
	25kV AC Overhead Line Installations – An Overview	121
	750v DC Third Rail Installations	122
	Civil Engineering	123
	Bridges, Viaducts and Tunnels	126
	Viaducts	128
	Tunnels	129
	Level Crossings	131
	Accommodation Crossings	133
	Other Civil Engineering Features	133
	Retaining Walls	134
Chapter 5	Railway Signalling	135
	Railway Signalling: A potted history from 1830 to the 1990s	135
	Signalling 101: Basic Principles	138
	Semaphore Signalling	138
	Semaphore Signal Types	139
	Colour Light Signalling	140
	Reading and Understanding Prototype Signalling Layouts	143
	Siting and Control of N Gauge Signals	145
	Signalling for Model Railways	145
	The Placement of Signals	146
	Plain Block Working	146
	Junction Working	147
	Refuges, Loops and Lie-Bys	148
	Signalling Within Station Limits	148
	Signalling for Termini	149
	Placement of Colour Light Signals	149
	Signal Sighting	150
	Siting and Modelling of Ancillary Signalling Equipment	151
	Train Protection Systems	152
	Automatic Warning System	152
	Automatic Train Protection	153
Chapter 6	Layout Design	154
	Layout Design	154
	The Nature of Trains	154
	Stations	155
	Loops and Refuges	159
	Sidings and Yards	160
	Maintenance and Servicing Depots	164

Chapter 7	Dressing the Scene	166
	Trackside Clutter	166
	Discarded Rails and Sleepers	167
	Cable Drums	167
	Ballast Bins	167
	Signs and Notices	167
	Industrial Railways	167
	Railway Buildings	170
	Maintenance Depots and Works	172
	Life Outside the Railway Fence	172
	The Scenic Break	173
	Vehicles	174
	BR Road Vehicles	175
	People	175
	Passengers	175
	Maintenance of Way Staff	176
	Depot and Yard Staff	179
Chapter 8	Realistic Operations	180
	Movement with a Purpose	180
	Prototypical Movements	180
	Correct Use of The Train Horn	181
	Train Behaviour at Level Crossings	182
	Basic Operations	182
	Running Round of Passenger Trains at Stations	183
	Carriage Sidings	185
	Shunting	185
	Block Working Using Multiple Operators	187
	Worked Example	187
	Developing a Working Timetable	187
	The Use of Wagon Cards	188
	The Utilisation of Randomised Task/Event Cards	190
Appendix A	TOPS Codes for Rolling Stock	191
Appendix B	Freight Stock Markings	198
Appendix C	Loads Permitted with Specific Brake Forces	199
Appendix D	Locomotive Life by Class	201
Appendix E	Block Signalling Bell Codes	202
Bibliography		204

Acknowledgements

I am thankful to the people who have provided information for the book and made comments on my manuscript, in particular my father, a railwayman of some 25 years' experience.

I am also indebted to the individuals who have contributed photographs for inclusion in the book, in particular David Ford, the late John Ford, Dave Peachey, Paul James and David Flitcroft. I hope that these important historical records reinforce the content of my text and provide examples and inspiration for the modeller. I also wish to thank Duncan Hunnisett, whose stunning layout 'Shirebrook' showcases the results possible if you carefully study the prototype.

Finally, I'd like to thank my partner Hayley, whose support during the many months spent drafting this work is greatly appreciated.

Preface

Compared to any other field of scale modelling, model railways offer perhaps the greatest opportunity to simulate realistic operation. Indeed, it is possible for an individual to achieve a realistically operated layout with only average modelling skills. What is required is an attention to detail, significant research and adherence (with some exceptions) to how the real railway operates.

This book is designed to give an overview of many aspects that contribute to the impression of a 'realistic' model railway. It covers train formations, signalling, civil engineering, track works and operating practice. The focus is on British 'N gauge' (1:148 scale), however, many, if not most, of the principles and advice given in this book have application across many layouts, regardless of scale.

It is, of course, necessary to emphasise that in this book, whilst drawing heavily on what went on in the real railway in years gone by, the focus is very much on the applicability to model. There are also operations that make perfect sense on a model railway, such as the 'fiddle yard', which is in effect an abstract representation of the remainder of the network not represented on the model, but whose existence and purpose on a real railway would be baffling and the operation of which would be totally impractical.

The Scope of the Book

To achieve what I wanted to achieve in a book this size is a huge task, perhaps approaching the impossible. What I lay out in the following pages is only really an introduction to the subject; it is impossible to cover all the various aspects associated with realistic railway operations in sufficient depth in a single volume. Therefore, think of this book more as an introduction to the subject of realistic operations, through which the railway modeller may start to refine the operation of their model railway layout.

I am keenly aware that railway modelling is a broad church, and that there is room in the hobby for everything from the simple bare board tailchaser to a minutely detailed and painstakingly researched reproduction of a prototypical railway environment. What interests one individual may not interest another, and there is absolutely nothing wrong with that.

This book is focused on British Railway practice, more specifically the British Rail TOPS Corporate (BR Blue) and Sectorisation eras, from the mid-1970s until the mid-1990s when shadow companies were formed in the run up to privatisation. The target is the beginner and intermediate model who, having perhaps built a 'train set' type layout in the past, wishes to design and develop a more realistic layout and operate it in a prototypically faithful manner.

Richard Middleton
July 2024

Chapter 1
The Railway As a System – Purpose, Destination, Traffic, Goods, Region and Geography

The real railway moves with a purpose, the transportation of people and goods between locations. The railways were originally built using funds from private investors and were constructed to turn a profit; they were not built for the enjoyment and interest of casual observers, a fact that seems to elude some enthusiasts.

As the railways developed, they quickly became the dominant mode of transport, and it was not until the widespread adoption of the motor car that this began to change. Over the years, the railway devised methods of operation that enabled them to serve their customers in a safe and efficient manner. Safety has always been at the forefront of railway operations, and great emphasis has always been placed on working within the varying rules and regulations.

It is a tendency of enthusiasts to glamourise and romanticise the railway environment, but it should not be forgotten that the lifeblood of the railway was not the crack express or the steam special, but the everyday mundane; the coal trains, the hoppers filled with ballast for the repair of the permanent way, and the noisy and unreliable two-car DMU plying the local route providing a vital link between communities and so on.

When we create a model railway, we need to try to embed this every day 'sense of purpose' from the real railway into our model. Moving trains aimlessly around a layout may be enjoyable, but it doesn't reflect what the real railway did. There is much more focus, especially on forums and in books, on absolute accuracy with regards to locomotives, stock, buildings etc, but far fewer modellers in general pay attention to prototypical accuracy and replicating in miniature how the real railway operated.

Operating your layout in a realistic manner is not only more authentic, but it can also be very enjoyable. It gives a purpose to the movement of every train on the layout and, if it involves co-operation between more than one operator, involves teamwork and good communication which can be immensely satisfying.

Operating a model in accordance with prototype practice can also be educational and whilst this may be regarded as an irrelevance for a home layout, it takes on a whole new meaning for an exhibition layout. By operating the layout in accordance with prototype practice, and displaying the timetable, the modeller is able to educate the wider public regarding the nature of railway operations and why things are the way they are.

Realistic operation is achieved by observation of numerous considerations, a prototypically correct track layout, correct placement of signals, the correct formation of trains, realistic civil engineering, and layout 'clutter'. These all add to the overall atmosphere and setting of the railway, tying it to a time and place, and thus making the whole ensemble more authentic.

Nothing in this text suggests that one should not build, operate, and enjoy a model railway that is wholly unrealistic; what gives one individual pleasure may not hold true for another. However, as you are reading this book, it shall be assumed that you wish to strive towards a realistic representation of a railway, within the confines of model engineering and your own skills and interests.

The BR Blue and Sectorisation Eras

The term 'modern image modelling' will be familiar to those who want to recreate the days when locomotives wearing Corporate Blue and Sectorisation liveries plied their trade on British Rail tracks.

The periods with which this book deals span an approximate timeframe between 1975 and 1994. This is the period in history when British Rail perhaps came the closest to operating a railway as a single integrated entity, and then proceeded to split its operations into business units, which was a foretaste of the privatised railway that was on the horizon.

Setting the Scene

Before constructing your model railway, you should attempt to 'set the scene'. What do I mean by this? Well, imagine you are creating a model of a railway set in the Midlands in the 1980s. You would seek inspiration from the railway of the period, and perhaps use the same or similar locomotives and stock. However, if you are going to create a fictitious railway, it makes sense to create a story and thus place the railway in the context of the real railway of the period.

Of course, this is much easier if you have selected a real location as the basis for your model. However, even with a real location, you may want to change certain aspects, or run trains that were never seen on the line in reality; therefore, it makes sense again to create a rationale why things are why they are on your layout.

It is also important to clearly define the railway network outside of your model. If you are creating an industrial railway, the 'network' outside of your model is the rest of the railway system. If you are creating a small suburban station set in the south west, then you may wish to consider the rest of the world which may begin/end at the boundaries of your model.

I find it useful to write a short history of the railway, perhaps one or two sides of A4. This will enable you to frame your layout against the historical precedents outlined in your history. If your layout is intended for exhibition, such a document can provide useful information to members of the public.

Reality versus Fantasy

As part of 'setting the scene' we must address the most important principle of realistic model railway operation, namely the differences between a real full-sized railway and that of a model. The real railway has a specific function, i.e., to move people and goods from one place to another. The function of the model, is, in most cases, to provide pleasure and interest for the owner (and audience if the layout is exhibited). Clearly, notwithstanding that the function of this volume is to inform the reader as to the considerations required to operate a model in a realistic manner, the fact remains that most modellers, even the 'rivet counters', enjoy an element of 'fun' on occasion. It is a human tendency to gravitate towards the 'bizarre' and model railways are no different. Weird and wonderful prototype locomotives, unusual liveries and uncommon traffic and workings are all features of model railways, just possibly at a higher frequency than occurred in real life!

Ultimately, there is of course, as with most things in life, a balance to be had. Just because something happened once on the real railway does not mean it should become a regular trope on a model. This applies in particular to painstakingly accurate models of prototype locations. It is a basic and indisputable fact; if you construct a faithfully accurate model of a location where very little happened in real life and then operate it strictly in accordance with prototype practice, then it should not be a surprise if nothing happens on the model!

After all, there is a reason why groups of trainspotters don't generally congregate on platforms of a rural halt; they want to be where the action is.

Therefore, the enthusiast should strive to build an operationally interesting layout, with due consideration for various regulations, practices and physical and operational limitations inherent to the real railway system. The rest of the chapters within this book attempt, in some small way, to inform the reader how this might be achieved, but ultimately, we are talking about 'your railway' and therefore, your rules should apply!

Chapter 2
Train Formations

One of the most straightforward methods by which to improve the realism of a model railway is with regard to train formations. Trains on the real railway are not assembled randomly, indeed they are put together meticulously, taking a number of factors and parameters into account. As with many things in this era, the composition of trains varied as the years progressed, and different operating areas could always be relied upon to produce regional variations. However, the basic rules outlined herein form a good basis for the assembly of realistic train formations for the average modeller. If absolute fidelity is required, then a review of Passenger Train Marshalling booklets, working timetables and photographs from the desired year/location can prove useful, as well as specialist reference volumes on the subject.

It is possible to devote entire volumes to the subject of train formations, and in a book of this size, it is only possible to give a flavour. This chapter gives an overview of factors affecting train formations, the TOPS system, various technical parameters, train lighting and headcodes, train classifications and vehicle compatibility. It also includes a brief overview of passenger, freight, departmental and special workings that an aspiring modeller may want to research further. Real world formations are provided in the form of tables, with headcodes, origin, destination and locomotive type included if the information is known.

Total Operations Processing System (TOPS) and Vehicle CARKND

One of the major issues plaguing the steam age railway was stock utilisation, i.e., maximising the usage of vehicles, the principle being that a vehicle not in traffic is a vehicle not earning any money. Traditionally, once a train had been discharged of its load, the 'empties' would be taken back to the starting point. Often this was because the vehicles' private companies owned them, but, and perhaps more significantly, it was the only real way that the management knew where the wagons were. It was clear that a more efficient method by which to utilise stock was needed, and this came in the form of Total Operations Processing System, or TOPS, introduced to British Rail in 1973.

TOPS was a computerised system introduced primarily to facilitate the movement of freight on the network but was also used for passenger stock and locomotives. TOPS was far more than a system for keeping records, it monitored and tracked the complete movement of vehicles from one destination to another. Thus, the location of each individual vehicle on the network could always be quickly ascertained.

Information was stored centrally on the TOPS computer database and there were terminals in all major freight centres and depots throughout the UK which allowed personnel to access the system.

The database stored all of the vehicle details as well as the location of wagons, for example, wagons undergoing maintenance or in long term storage. Timetables indicated the likely arrival of a particular vehicle when it was in transit.

Operating parameters of each vehicle were stored within the database to enable each vehicle to be easily assigned to a particular task. This would include the vehicle number, technical particulars, braking system, and tare (empty) and gross laden weight (GLW).

The most visible aspect of TOPS was the panel on the side of freight stock (different panels were located on passenger stock and locomotives).

How to Understand Freight Vehicle TOPS Codes

Starting from the introduction of TOPS, all vehicles were assigned a TOPS code. However, for operational reasons, only wagons, departmental vehicles and some on-track plant displayed the code prominently. Coaches displayed a differently formatted code on the body ends, but in N gauge this panel is generally too small to be read, Therefore, freight wagons, being the vehicles where TOPS codes provide the greatest amount of compatibility information to the modeller, and thus assist in accurate train formation, are the primary focus of this section.

What we as modellers normally call the 'TOPS code' on the side of the wagon was actually called the CARKND on the TOPS system. This three-letter code, sitting inside a data panel which also contained a unique vehicle number, tare and laden weights and sometimes other information, can reveal a surprising amount of information relating to the vehicle, if we are able to decipher it. A basic description of the code system is given below.

The first letter of the code describes the vehicle category. Let us imagine we have seen a wagon in a siding with the CARKND 'ZKV'. If we consult the table in Appendix A we discover that the letter 'Z' denotes a twin-axle departmental wagon. The second letter

Body End TOPS Data Panel on Mk 3 Coach, the TOPS Code is the 4-letter code in the Top Left Corner, in this example AD1H is the TOPS code for a Mark 3 First Open (FO). Coaching stock was not added to TOPS until 1983.

gives a more detailed explanation of what type of vehicle we are looking at. Again, if we refer to the table we find that that 'K' denotes a Ironstone Hopper with a tare weight of over 26 tonnes. Finally, the last letter in the sequence tells us what type of brake system is fitted to this vehicle. In this case the letter is 'V' which we find from Appendix A means that the vehicle is fitted with vacuum brakes. Thus, we now understand that the wagon we are looking at is a twin-axle ex-ironstone wagon in departmental service which is equipped with vacuum brakes. Using this basic approach, it becomes straightforward to identify what the vehicles are, based on their TOPS code. However, from the perspective of realistic train formations, the most important part of the code is the brake code, because this tells us what brake system(s) are fitted to the vehicle in question, and thus can have a significant impact on train formations and vehicle compatibility.

An ex-Ironstone Wagon, TOPS Code ZKV, in use by the engineers photographed in Hereford Yard in 1989. (*Jamerail/Flickr*)

Broadly speaking, the following rules applied:

- Air braked stock was only compatible with other air braked stock
- Vacuum braked stock was only compatible with other vacuum braked stock
- A vacuum braked vehicle through piped for air could operate in an air braked train, but its vacuum brakes wouldn't work
- An air piped vehicle through piped for vacuum could operate in a vacuum braked train, but its air brakes wouldn't work.
- Dual braked vehicles could operate in any train.
- Vacuum braked wagons were often marshalled next to the locomotive in an otherwise completely unfitted freight train. This was referred to as a 'fitted head' and allowed the train to run at a higher maximum speed.
- All unfitted trains had at least one brakevan, normally, but not always, marshalled at the end of the train.
- Fully unfitted trains had been eliminated by 1985; although the engineers continued to use unfitted departmental wagons into the 1990s.

It was common in the early TOPS era for vehicles not to display a code simply because no one had got round to painting one on. Similarly, it wasn't unusual to find codes crudely hand painted on the sides of a particularly dilapidated vehicle, the approach being either to wait until the vehicle was repainted for the application of a 'proper' code, or not to bother as the vehicle was near the end of its life and thus close to being scrapped.

Vehicle Compatibility – Couplers, Jumpers, Barrier, and Translator Vehicles

On the real railway, it is not quite as simple as just connecting two vehicles together and driving off. There are a number of interfaces between two vehicles. Such connections, as well as providing a mechanical link between two rail vehicles to allow them to be formed into a train, provide electrical power for train control, power for passenger vehicles, and air or vacuum connections for braking systems.

From the 1970s onwards British Rail introduced many new types of vehicles, some of which moved away from the traditional drawhook and screw-coupler style of coupling. This resulted in a plethora of different coupler types, and vehicles that had a wide variety of mechanical and electrical compatibility with one another.

Therefore, in order to model realistic operations in this time frame, a modeller must be cognisant of rolling stock compatibility and what vehicles could couple to and operate with one another.

Before the 1970s, most locomotives used a drawhook and screwlink coupling system. Wagon couplings were either an 'Instanter' coupler or an adjustable screw coupler.

Several types of multiple units, some locomotives and coaching stock were equipped with 'drophead' couplers which incorporated a standard drawhook and a buckeye coupling which allowed for semi-automatic mechanical coupling. Similar 'buckeye' couplers were also used in a limited manner on certain wagon fleets, particularly tipplers (a special rotating drawgear was fitted to allow the wagons to remain coupled together when tipped).

These three coupling systems provided only mechanical coupling, the air or vacuum hoses for braking and electrical jumper cables for control of other vehicles or the supply of electric train heating to coaches had to be connected separately. However, in most respects, vehicles equipped with this type of coupler were largely compatible with one another, and it was theoretically possible to haul a vehicle with another vehicle even if they weren't electrically or pneumatically compatible.

In the 1970s and 1980s, the introduction of the several types of Electric and Diesel Multiple Unit introduced several new types of coupler. These usually provided automatic coupling, and electrical and pneumatic connections all in one unit, but were largely incompatible with existing stock. These couplers included the 'Tightlock' type fitted mostly to EMUs and

Drophead automatic coupler in the dropped position, showing the standard drawhook.

Coupling and brake connections on the headstock of a dual braked Class 47 locomotive. The thin hoses are for the air braked system, the larger diameter hoses are for the vacuum braking system. The ribbed hose carries steam for heating coaching stock. (*Jamerail/Flickr*)

EMU Translator Vans such as this T1 set, were converted from redundant passenger stock by Network SouthEast in the late 1980s. The vehicles are fitted with standard couplers one end and tightlock couplers the other. The interior of the vehicles houses the translation equipment, which provides compatibility between the 'Westcode' brake system of the EMU and the standard brake system of the locomotive.

Mk 4 coaches, and the BSI Lightweight Coupler fitted to 2nd Generation DMUs.

A number of special vehicles were introduced to enable these vehicles to be hauled by existing traction to and from main works for overhaul attention. A barrier or match vehicle is simply a vehicle that provides mechanical compatibility between the couplers in the train. A translator vehicle takes this one step further and allows a locomotive to operate an otherwise incompatible braking system used on the vehicles being hauled. Two vans are normally provided, one at each end of the train.

Barrier vehicles ranged from simple local conversions of redundant wagons or chassis to full conversions of former passenger stock. In some cases, most usually for emergency use or rescue and recovery situations, special 'adaptor' couplings were provided. The use of such adaptors normally meant operational limitations such as a speed restriction.

Translator vehicles were created using redundant coaching stock, and were used primarily in Southern Region who were the largest operator of Electric Multiple Units.

When the High Speed Train (HST) was introduced in 1975, it became apparent that a method was required to haul individual HST trailers between locations for maintenance and operational purposes using standard locomotives. The Mk 3 HST trailer was fitted with a fixed buckeye coupler and could not be coupled to a standard locomotive coupler. A number of redundant Mk 1 and Mk 2 coaches were converted to 'HST Barrier Vehicles', this typically consisted of the windows being plated over, the interiors stripped out and an Alliance coupler being fitted to one end of the coach. This allowed any locomotive to haul the trailer vehicles in a similar method to the HST power cars.

London Underground stock used a coupler known as a 'wedgelock' which was incompatible with all BR stock and required a uniquely equipped interface wagon to be coupled between the locomotive and the underground vehicles.

Match Wagon in use to haul Central Line stock for attention at Derby Litchurch Lane Works, 1989. (*Dave Peachey*)

Match wagons, translator vehicles and barrier vans can easily be represented by models with the standard N Gauge coupler used, as long as the livery and markings of such vehicles are replicated in order to inform the observer as to the purpose of the vehicle within the formation.

Train Marker Lighting, Lamps and Headcodes

Headcodes

At the time of the British Railways Modernisation Plan in 1955, steam locomotives displayed route headcode discs or white lamps to help signallers identify different trains. The first diesel locomotives were fitted with four position headcode discs – flapped discs attached to the locomotive front which showed white or a light when open. Varying combinations of open and closed discs denoted different types of train. The disc system fell out of use by 1960 but locomotives continued to carry the discs for many years afterwards, with locomotives fitted with headcode discs present until at least the 1980s. The displayed 'code' therefore bore little or no resemblance to the actual train!

In June 1961, with the intended removal of steam locomotives and the replacement of manual signalboxes with centralised control power signal boxes (PSBs) BR introduced a four-letter train reporting code, referred to as a 'headcode'. The code was displayed on the front of the locomotive and on panel displays in signal boxes. By the late 1970s, locomotives no longer displayed the code, but they continued to be used in signal boxes and on working timetables and were referred to as a train reporting number.

The basis of the system is that all trains are allocated a four-character code which are never duplicated over one section of line. The code is constructed thus:

First Letter	- Class of Train	- Numbers 0 through 9
Second Letter (Excluding I, Q, R U, W and Y)	- Destination District	- Letters of the Alphabet
Third/Fourth Letter	- Individual Train Numbers	- Numbers 0 through 9

NOTE: Local and Passenger and ECS Trains were denoted by a ROUTE number. Freight 'trip' workings were denoted by a TRIP number.

The first letter shows the Class of train, typically this was as below:

1992 BR Train Class Table

Code	Train Class
1	Express Passenger, Breakdown or Snowplough Duty
2	Ordinary; Suburban or branch passenger or mixed. Breakdown train not on duty.
3	Parcels or perishable goods train composed of bogie fitted vehicles
4	Freight train timed at 75mph
5	Empty Carriage Stock (ECS)
6	Freight train timed at over 50mph
7	Freight train timed at less than 45mph
8	Freight train timed at less than 35mph
9	Train with some or all vehicles not fitted with continuous brakes
0	Light Locomotive

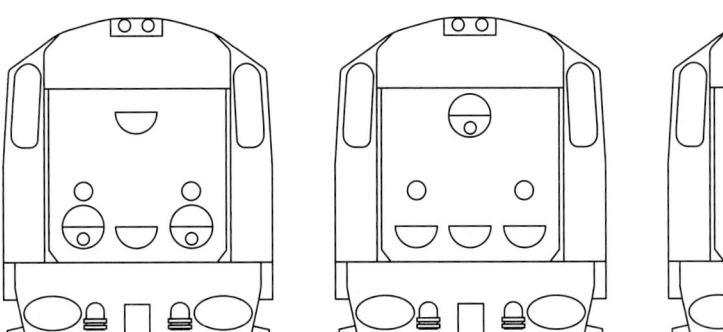

Some common disc headcode train classifications – from left to right – Express Passenger Train, Ordinary Stopping Passenger Train, Unfitted Freight Train, Empty Coaching Stock (ECS) or Parcels Train

The second letter indicates the destination of the train. For a train that started and ended within the same region, this would indicate a destination district. An example table gives some of the Destination District codes within the London Midland Region.

Code	Destination Area
A	Euston
B	Euston – Rugby
C	St Pancras – Marylebone – Manchester North
D	Chester – Nottingham
F	Leicester
G	Birmingham
H	Manchester South – Stoke-on-Trent
J	Manchester North
K	Crewe – Liverpool Lime Street – Liverpool Central
L	Preston (excluding Fylde) – Barrow – Carlisle
P	Blackpool and Fylde – Derby

For trains that originate and terminate in different regions, this would indicate the destination region. Destination regions for inter-regional trains are given in the table below.

Inter-Region Codes

Code	Destination Area
E	Eastern Region
M	London Midland Region
N	North Eastern Region
O	Southern Region
S	Scottish Region
V	Western Region
X	Excursion or Special

A trip working within a region would normally be assigned a number rather than a complete headcode, such as 'Trip 62'.

Train and Brake Van Lighting

After 1969, brake vans were only required to be provided for trains carrying dangerous goods, fully unfitted trains and for other special circumstances. Where provided, brake vans had to show a red lamp to the rear (to show the back of the train) and 'side lamps' in certain configurations as described below.

7.4 Side Lamps
7.4.1 Guards of freight trains not fitted throughout with the automatic brake must ensure that in addition to the train tail lamp, two side lamps are carried on the rearmost brakevan. After sunset or during fog or falling snow they must show a white light forward, but the indication to the rear must be as follows:

a) On main lines, fast lines and single lines – two red lights.
b) On slow lines, relief lines or loops adjoining main or fast lines and running in the same direction – one

red light on the side furthest away from the main or fast line and one white light on the side nearest the main or fast line.

c) On goods lines or loops adjoining slow or relief lines and running in the same direction – two red lights.

d). On reception sidings – the side lights must be removed or obscured when the train has passed into the sidings.

When a train was being propelled (i.e., a locomotive was pushing the train from the rear) the following rules generally applied:

a) Propelling in the right direction – one white light on the leading vehicle unless otherwise specially instructed

b) Propelling in the wrong direction – one red light on the leading vehicle unless otherwise specially instructed (see below)

c) When the front portion of a divided train is being set back onto the rear portion – one white light on leading vehicle.

d) A train that is proceeding from signal box in advance to assist a disabled train – one white light on leading vehicle.

In addition to the special lighting configurations required for brakevans, the last vehicle in any train (regardless of its type) was required to show a single red light to the rear. The purpose of this light was actually very simple, it provided a clear indication to the signalman, and any other interested parties, that the train was complete, i.e., wagons had not become

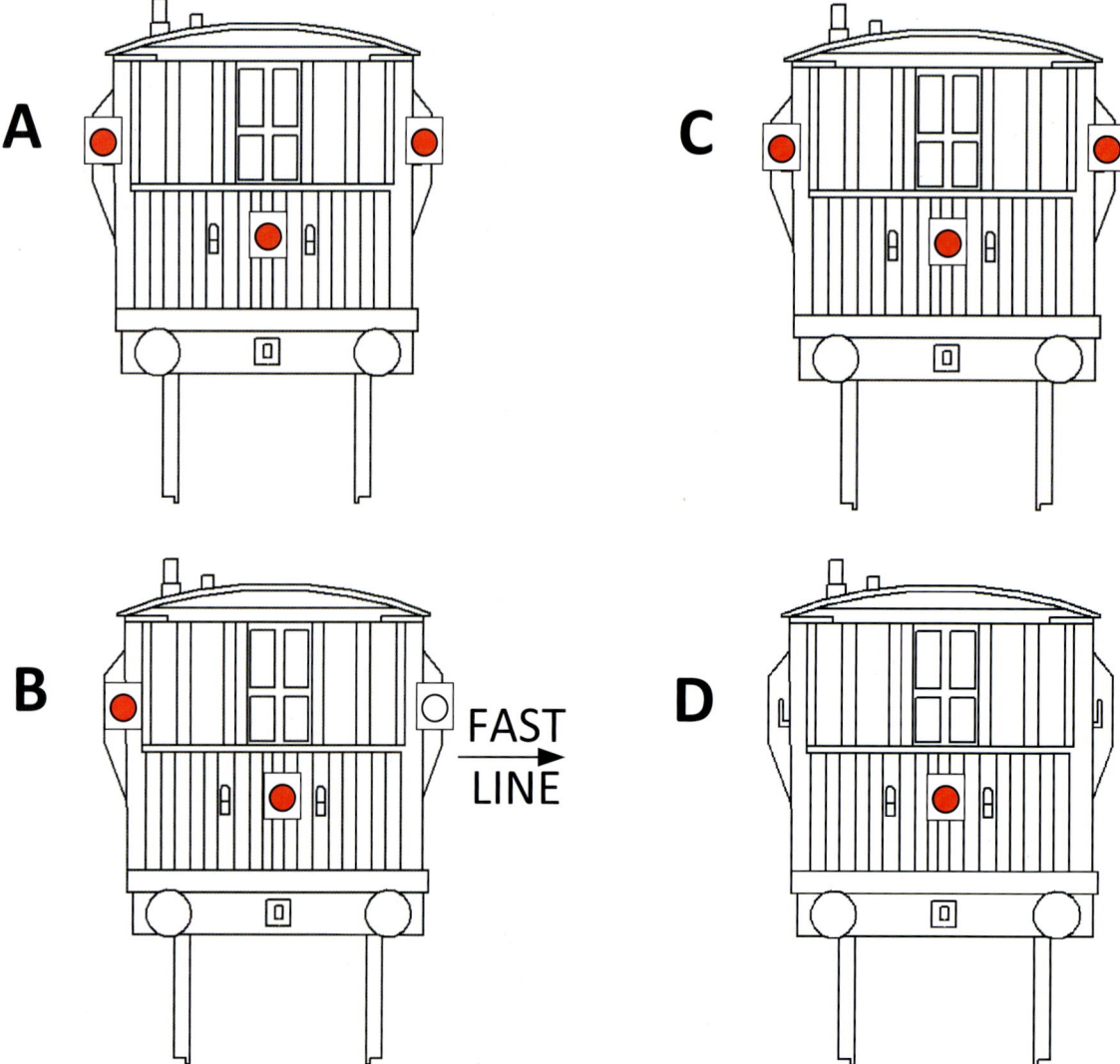

Brake van side lamp configurations. Note that the middle lamp is the 'train tail lamp' and would always be lit to signify the end of the train.

detached from the train during transit. This was particularly important when unfitted freight trains were still common, as a wagon becoming detached from a fitted train would drop the air (or destroy the vacuum) and bring the train safely to a halt.

Multiple units and locomotives had their own integrated lighting systems, and the red lamp to the rear would be replaced by a pair of red rear marker lights. On a two-cab unit such as a locomotive or multiple unit, this is a useful indicator as to direction of travel when looking at photographs.

Selecting Appropriate Traction for Your Train

Another aspect of TOPS that became very useful was assigning the multitude of locomotives and multiple units in use at the time a Class number. A Class could be further sub-divided to encompass locomotive differences, for example those fitted with boilers or electric train supply heating, and those without.

Each mainline locomotive carried a standard data panel prominently stencilled in white or black (depending on livery) on the bodyside. The code, A, V or X indicated air, vacuum or dual braking. The Class, weight (always in working order, i.e., with fuel and water) and maximum speed was also included on the panel. The brake force was an indication of the locomotive's stopping power. It is shown in tonnes and calculated via a formula contained in the working instructions which depended on the locomotive weight and other factors. Locomotives equipped with Electric Train Heating displayed an ETH index on their panel; this indicates the power output of the ETH system fitted to the locomotive.

The Route Availability (RA) number was based on the weight on each axle on the locomotive, no locomotive was permitted to operate across a route with a lower RA number than was shown on the panel.

If you are modelling a fictional location, it is not important to consider Route Availability, however, those modelling prototypical locations (or those based on such locations), should factor in the locomotive's Route Availability. Generally, locomotives with a Route Availability of RA 7 or below had access to most of the network. Locomotives with a Route Availability of RA 5 had access to nearly all routes (including lightly used lines such as the West Highland Line). Locomotives with an availability of RA 1 or 2 were typically assigned to duties involving the navigation of very tight radius curves, such as those found in docks and yards. Locomotives with a rating above 7 were typically restricted to specific operating areas, and rarely strayed from these areas.

A useful publication if it can be found is British Rail document BR 29993 'Route Availability of Diesel and Electric Locomotives'. Although covering the Eastern region only it contains detailed maps for each line showing what locomotives could operate where; similar publications were presumably produced for other regions of BR.

In a similar manner to locomotives, wagons also had a Route Availability rating, and the highest rated vehicle in a train (including the locomotive) gave the train its assigned RA.

Each specific wagon type had its own RA rating, but in an emergency, the BR Rule Book gave the following values for different types of vehicles.

Two Axle Vehicles	RA
Weight Per Axle (Tonnes)	
13 or less	1
13 – 15.2	2
15.2 – 16.3	3
16.3 – 17.3	4
17.3 – 18.3	5
18.3 – 20.3	6
20.3 – 21.3	7
21.3 – 22.9	8
22.9 – 23.4	9
23.4 – 25.4	10
Three Axle Vehicles	**RA**
Milk Tanks	1
Motor Car Carrying (Articulated) (Private Owner)	2
All others	1

Vehicle Type	Brake Force (Tons)
SPA Plate Wagon	27 Laden / 7 Empty
STV Tube Wagon	6
TTA 41t Tank Wagon	22 Laden / 9 Empty
YBA Sturgeon	21
YCV Turbot	9
YGB Seacow Hopper	31 Laden / 11 Empty
YGH Sealion Hopper	21 Laden / 11 Empty
YMA Salmon	22
VBA Van	27 Laden / 12 Empty
VCA Van	27 Laden / 12 Empty
VDA Van	21 Laden / 7 Empty
VEA Van	11 Laden / 4 Empty
VGA Van	45 Laden / 10 Empty

A Word on Uniformity

One of the tendencies of many modellers is the insistence upon of uniform rakes of coaches or wagons, i.e., a rake of vehicles that are either all of the same type or, just as commonly, all painted in the same livery. This is probably because uniform trains present a pleasing visual image and just 'look right'.

Of course, this happened on the real railway, for example, by the mid-1970s all passenger coaching stock was in BR Blue and Grey, but, in the case of passenger stock, as the 1980s progressed and the new Intercity and Provincial Liveries began to appear, mixed rakes, both of coach types (for example Mk 1 and Mk 2 coaching stock often operated together in the same train) and liveries, became common. This state of affairs continued well into the 1990s, especially on services where 'scratch' rakes of coaching stock were put together, such as seasonal holiday trains.

Wagons are even more of a mixed bag and the traditional mixed freight and the later Speedlink services give the modeller a template to allow many types of freight vehicle in the same train.

Even the ubiquitous 'uniform' MGR train had visual variations. Seen here approaching the site of the former Stanton gate station, 58013 and 56108 haul MGR trains away from Toton in 1985. Note the contrast between the recently shopped and repainted HAA hoppers and the well-used and weathered examples that have been in service for some time. (*Dave Peachey*)

detached from the train during transit. This was particularly important when unfitted freight trains were still common, as a wagon becoming detached from a fitted train would drop the air (or destroy the vacuum) and bring the train safely to a halt.

Multiple units and locomotives had their own integrated lighting systems, and the red lamp to the rear would be replaced by a pair of red rear marker lights. On a two-cab unit such as a locomotive or multiple unit, this is a useful indicator as to direction of travel when looking at photographs.

Selecting Appropriate Traction for Your Train

Another aspect of TOPS that became very useful was assigning the multitude of locomotives and multiple units in use at the time a Class number. A Class could be further sub-divided to encompass locomotive differences, for example those fitted with boilers or electric train supply heating, and those without.

Each mainline locomotive carried a standard data panel prominently stencilled in white or black (depending on livery) on the bodyside. The code, A, V or X indicated air, vacuum or dual braking. The Class, weight (always in working order, i.e., with fuel and water) and maximum speed was also included on the panel. The brake force was an indication of the locomotive's stopping power. It is shown in tonnes and calculated via a formula contained in the working instructions which depended on the locomotive weight and other factors. Locomotives equipped with Electric Train Heating displayed an ETH index on their panel; this indicates the power output of the ETH system fitted to the locomotive.

The Route Availability (RA) number was based on the weight on each axle on the locomotive, no locomotive was permitted to operate across a route with a lower RA number than was shown on the panel.

If you are modelling a fictional location, it is not important to consider Route Availability, however, those modelling prototypical locations (or those based on such locations), should factor in the locomotive's Route Availability. Generally, locomotives with a Route Availability of RA 7 or below had access to most of the network. Locomotives with a Route Availability of RA 5 had access to nearly all routes (including lightly used lines such as the West Highland Line). Locomotives with an availability of RA 1 or 2 were typically assigned to duties involving the navigation of very tight radius curves, such as those found in docks and yards. Locomotives with a rating above 7 were typically restricted to specific operating areas, and rarely strayed from these areas.

A useful publication if it can be found is British Rail document BR 29993 'Route Availability of Diesel and Electric Locomotives'. Although covering the Eastern region only it contains detailed maps for each line showing what locomotives could operate where; similar publications were presumably produced for other regions of BR.

In a similar manner to locomotives, wagons also had a Route Availability rating, and the highest rated vehicle in a train (including the locomotive) gave the train its assigned RA.

Each specific wagon type had its own RA rating, but in an emergency, the BR Rule Book gave the following values for different types of vehicles.

Two Axle Vehicles	RA
Weight Per Axle (Tonnes)	
13 or less	1
13 – 15.2	2
15.2 – 16.3	3
16.3 – 17.3	4
17.3 – 18.3	5
18.3 – 20.3	6
20.3 – 21.3	7
21.3 – 22.9	8
22.9 – 23.4	9
23.4 – 25.4	10
Three Axle Vehicles	**RA**
Milk Tanks	1
Motor Car Carrying (Articulated) (Private Owner)	2
All others	1

Four Axle Vehicles	RA
Weight Per Axle (Tonnes)	
13 or less	3
13 – 14	4
14 – 15.2	5
15.2 – 16.3	6
16.3 – 17.3	7
All Coaching Stock (Including NPCCS)	2

Other Compatibility Considerations – ETH Index and Train Brake Force

Each coach equipped with Electric Train Heating (ETH) had a rating as shown in the table below. This was also displayed on the coach's data panel.

The total of all ETH equipped coaches in a train could not exceed the ETH rating of the locomotive, as the electric train supply would not function, or, in some cases, the locomotive would not have enough power to operate.

ETH Ratings for Various BR Stock as of 1985

Type of Coach	ETH Index
Mk 1 BG, RKB	1
Mk 1 RBR Mk 1 BFK, BSK	2
Newspaper Van Mk 1 BSO TPO Mk 1 BCK Mk 1 RMB Mk 1 FO, FK, CK	3
Mk 1 SK, SO, TSO PFB Mk 2, 2a, and 2c BSO Mk 2, 2a, 2b, 2c SO, FK, SK, CK, BFK, TSO	4
PFP Mk 2d, 2e, 2f SO, FO, BSO, BFK, FK, TSO	5

Type of Coach	ETH Index
PFK Mk 3a TSO, FO, SLE, SLEP (tea boiler not in use) Mk 3b FO, BFO	6
Mk 3a SLEP (tea boiler in use)	7
Mk 3a RFB, RSM	14
Class 491 Stock (4 car set)	20
Class 488/2 (two-car set)	10
Class 488/3 (three-car set)	15

Take as an example a Class 31/4 locomotive. This locomotive has an ETH index of 66. We have a train consisting of 9 Mk 2a SO, 2x Mk 2a FO, 1x Mk 1 RMB and 1x Mk 2a BSO. Using the ETH figures given in the table above this would give us a total train index of 52, meaning that the locomotive can theoretically supply sufficient power to the train. Unfortunately, in real life the more coaches that a Class 31/4 locomotive had to haul, the less power available for traction due to the power being taken directly from the locomotives traction supply. Using this type of locomotive with a train close to its maximum ETH rating resulted in there being around a third less power available than the same locomotive with no ETH draw. This meant in practice trains hauled by this type of locomotive rarely exceeded five coaches in length.

The ratings of the most common types of equipped locomotives are given in the table below.

Class	ETH Index	Notes
31/4	66	
33	48	
37/4	30	
45/1	66	
47/4	75	
47/7	66	
50	61	Air-Conditioned Mk 2 Stock must be pre-conditioned. If this cannot be provided the index is limited to 48.

Class	ETH Index	Notes
73, 81, 82, 83, 85, 86	66	These locomotives have an Index of 75 if all the coaches in the train have the Suffix 'X' after their ETH rating
87	66	These locomotives have an Index of 95 if all the coaches in the train have the Suffix 'X' after their ETH rating
489	66	

Another consideration was train weight and required train brake forces. Detailed tables were provided in the Working Manual for Rail Staff to enable groundstaff and traincrew to accurately determine train weights. If this was ignored or calculated incorrectly, particularly for trains without a continuous brake, it could have catastrophic results.

Working out the brakeforce of a train could be tricky and would normally be calculated on the TOPS computer and supplied to the traincrew in the TOPS trainlist. It depended on a number of factors, including the route (most specifically gradients) and the maximum permitted speed of the train.

As an example, let us assume we have a train comprising 32 empty HAA Coal Hoppers hauled by a Class 56 locomotive. Each coal hopper weighs 14 tons empty, giving us a total train weight of 448 tons (locomotives are not counted unless being hauled 'dead'). The maximum speed of an unladen HAA hopper was 45mph.

The Class 56 locomotive has a brake force of 60 tons. Each coal hopper has a brake force of 10 tons. Therefore, the entire train has a brake force of 380 tons. If we refer to the brake force table provided in Appendix C we find that a train with a brake force of 380 tons operating at 45mph can haul a train with a weight of up to 1810 tons, therefore our train, only weighing 448 tons, can easily operate at the required speed.

Using this method, the modeller can realistically calculate whether their train has the required brake force to operate safely at the maximum train speed.

Note that only wagons and locomotives fitted with **operational** automatic brakes (air or vacuum) count towards the brake force rating. However, the weight of all vehicles in the train, even dead/unfitted ones count towards the total train weight.

The brake force weights of some common wagons are given in the table below.

Vehicle Type	Brake Force (Tons)
BAA Bogie Steel Carrier	50 Laden / 14 Empty
BCV Bogie Bolster C	9
BDV Bogie Bolster D	12
BDA Bogie Bolster	40 Laden / 19 Empty
BPV Boplate	12
CDA China Clay Hopper	10
CHV Hopper	10 Laden / 5 Empty
HAA Coal Hopper	10
HEA Hopper	24 Laden / 8 Empty
HTV Hopper	6
MCV 16t Mineral	11 Laden / 4 Empty
MDV 24t Mineral	12 Laden / 6 Empty
MSV Tippler	15 Laden / 5 Empty
MTV Tipper	14 Laden / 7 Empty
NCV Mk 1 Brake Van (Newspapers)	17
NGV Carflat	11
NJA GUV	19.6
OBA Open Wagon	24 Laden / 7 Empty
OCA Open Wagon	24 Laden / 7 Empty
OOV China Clay Open	3
PCA Cement Wagon	27 Laden / 10 Empty
PCV 'Presflo'	12 Laden / 6 Empty
SPV Plate Wagon	6

Vehicle Type	Brake Force (Tons)
SPA Plate Wagon	27 Laden / 7 Empty
STV Tube Wagon	6
TTA 41t Tank Wagon	22 Laden / 9 Empty
YBA Sturgeon	21
YCV Turbot	9
YGB Seacow Hopper	31 Laden / 11 Empty
YGH Sealion Hopper	21 Laden / 11 Empty
YMA Salmon	22
VBA Van	27 Laden / 12 Empty
VCA Van	27 Laden / 12 Empty
VDA Van	21 Laden / 7 Empty
VEA Van	11 Laden / 4 Empty
VGA Van	45 Laden / 10 Empty

A Word on Uniformity

One of the tendencies of many modellers is the insistence upon of uniform rakes of coaches or wagons, i.e., a rake of vehicles that are either all of the same type or, just as commonly, all painted in the same livery. This is probably because uniform trains present a pleasing visual image and just 'look right'.

Of course, this happened on the real railway, for example, by the mid-1970s all passenger coaching stock was in BR Blue and Grey, but, in the case of passenger stock, as the 1980s progressed and the new Intercity and Provincial Liveries began to appear, mixed rakes, both of coach types (for example Mk 1 and Mk 2 coaching stock often operated together in the same train) and liveries, became common. This state of affairs continued well into the 1990s, especially on services where 'scratch' rakes of coaching stock were put together, such as seasonal holiday trains.

Wagons are even more of a mixed bag and the traditional mixed freight and the later Speedlink services give the modeller a template to allow many types of freight vehicle in the same train.

Even the ubiquitous 'uniform' MGR train had visual variations. Seen here approaching the site of the former Stanton gate station, 58013 and 56108 haul MGR trains away from Toton in 1985. Note the contrast between the recently shopped and repainted HAA hoppers and the well-used and weathered examples that have been in service for some time. (*Dave Peachey*)

Even if the majority of the wagons are the same, it took many years for wagons to be repainted and some never received a new livery, so trains with mixed liveries were common.

Clearly uniform rakes of wagons became increasingly prevalent as the network moved from the traditional 'mixed freight' to block trains. However, even if we consider arguably the most common type of air-braked freight, the MGR Coal train, a variety of livery variations could be seen in a train comprised largely of identical vehicles.

In order to better reflect the railway of the time, the modeller should critically review their assembled train formations and consider whether they are 'too uniform'. As ever, reference to period photographs will yield the most accurate results.

Passenger Train Formations

A feature of passenger services in this era was the rise of the multiple unit train. This was physically realised by the introduction of a number of electric multiple units and second-generation diesel multiple units starting in the late 1970s and continuing through to the 1990s. The addition of the HST in 1975 saw a significant contraction in traditional locomotive hauled services.

Even with the introduction of multiple units and the HST in 1975, there were still a surprising number of locomotive hauled services operating well into the 1990s. In addition, shortages of multiple units in the 1980s resulted in many scratch 'DMU replacement' short-formed services comprised of a diesel locomotive (most often a Class 31) and three or four coaches. This section will focus mainly on locomotive hauled services, with some discussion of HST set formations.

As with all train formations, the exact composition of a train varied across regions and time periods. There were also, of course, occasional failures resulting in some unusual workings. However, the following text gives an outline of passenger operations in the period.

Intercity and Cross-Country workings

Longer-distance routes radiating outwards from London fell under the 'Intercity' heading, together with a group of workings known as 'Cross-Country' centred around Birmingham. Intercity and Cross-Country routes principally served the economically important routes, which had received significant investment over the years.

Intercity trains were the crack express trains of the day and included famous named trains

A Typical 1980s Intercity Working, with a Mixture of Mark 2 Coaches and a Mark 1 Buffet. This is the 12:15 Derby to London St Pancras Service, Seen Here Passing Loughborough. Note the lone Intercity liveried coach, this would become a regular sight as the decade progressed and more stock was repainted in Sectorisation colours. (*Dave Peachey*)

(such as the Master Cutler) in addition to the everyday trains that linked the country's major cities. The introduction of the HST in 1975 meant that principal routes gradually lost their locomotive hauled express trains, none more so keenly felt than the loss of the 'Deltic' hauled expresses on the East Coast Mainline in 1978. Initially taken over by HSTs, East Coast workings had reverted to modern locomotive hauled services by the end of the 1980s, utilising the new Class 91 Locomotives and Mk 4 Coaching Stock.

On the Western Region, the removal of the Diesel-Hydraulic locomotives in the mid-1970s left workings in the hands of Class 47 and Class 50 diesel locomotives. In the Midlands, the Midland Mainline used Class 45 and Class 47 locomotives until the early 1980s, when all Intercity workings were taken over by HSTs.

By 1984, Mk 1 coaching stock had been relegated to secondary workings, and locomotive hauled Intercity and Cross-Country services were served largely by Mk 2 and Mk 2 Air-Conditioned coaches, often mixed within the same train. A Mk 1 buffet or restaurant coach would often be included in the formation, as such coaches were not produced in Mk 2 guise.

Finally, locomotive hauled Mk 3 coaches were introduced from 1975, and used almost exclusively on the West Coast mainline, hauled mainly by electric locomotives.

St Pancras – Sheffield, 'The Master Cutler' May 1977	
Locomotive Number: 45124	
Vehicle Type	Load
MK 2D BSO	Passengers
MK 2D SO x3	Passengers
MK 1 RMB	Passengers
MK 2D FO x3	Passengers

Derby–St Pancras, April 1988	
Locomotive Number: 45128	
Vehicle Type	Load
Mk 2A SO	Passengers
MK 2A SO x2	Passengers

Derby–St Pancras, April 1988	
MK 1 RMB	Passengers
MK 2F FO	Passengers
MK 2A SO	Passengers
MK2A BSO	Passengers
MK2A SO	Passengers

High Speed Train

The HST was arguably the biggest success of the Corporate Blue Era and would be inextricably linked to modelling most mainline locations any time after the late 1970s.

When originally introduced it was intended that the train would remain as a complete operational set so as such they were given a unit style numbering sequence starting in 253/254xxx. It soon became apparent working the sets in this way was impractical, as a fault on a single coach or power car rendered the whole set unusable until the fault was rectified. Later the power cars were given the classification Class 43 under TOPS and became interchangeable between coaching stock rakes.

HST formations in the BR Blue and Sectorisation eras were largely fixed and varied mostly by region. It was common to see a mix of BR Blue/Executive and Executive/Swallow liveries during the cross-over period between each livery. The power cars tended to be repainted before the coaches.

A rough outline of the main HST formations upon introduction can be found below.

- Class 253 Western Region 2+7 – 2TF TRUK TS TRSB 2TS
- Class 254 Eastern/Scottish Region 2+8 – 2TF TRUK 2TS TRSB 2TS
- London Midland Region 2+7 – 2TF TRUB TS TRSB TGS TS

In 1980, complaints about the noise in the rear of the power cars led to the introduction of a TGS (Trailer Guard Second) which replaced one of the Second-Class Coaches. This coach had a compartment for the guard to sit in, and the practice of sitting in the rear of the power car was discontinued.

TRAIN FORMATIONS • 29

Set 253020 heads West at Speed through Maidenhead in 1981. (*Jamerail/Flickr*)

Set 254008 on the Down Fast at the Top of Holloway Bank, July 1980. (*Jamerail/Flickr*)

This atmospheric shot of a Midland Region HST passing Stenson Junction in 1991, shows the 6-car formation of the Midland sets. (*Dave Peachey*)

The TRUK cars were quickly replaced by a TS on the Western Region and most had been replaced on the Eastern Region by 1985. TRUB cars (Trailer Restaurant Unclassified Buffet) were built from 1978 to replace the TRUK cars, these were reclassified as TRFB (Trailer Restaurant First Buffet) from 1985 on the Eastern and London Midland Regions and from 1989/90 on the Western Region.

Later in their careers, modified coach designs were introduced and therefore the makeup of various sets changed. For fidelity to the time period and region, Passenger Train Marshalling documents relevant to the period and operating location should be consulted.

An unusual variation of the Class 43 power car was introduced in 1987. When the electrification of the East Coast Main Line was underway, BR realised that the new Mk 4 coaches then under construction were not going to be ready in time for the first electric hauled services on the route. A total of fourteen power cars were modified by having their front valances removed and buffers fitted. They also received remote control equipment so that they could work with the new Class 91 locomotive. The power cars worked the East Coast until 1991, when they had the control equipment removed prior to being returned to capital stock. The buffers were retained.

Of course, there were always unusual workings as a result of testing, failures etc. Lots of photographs can be found on the internet of unusual HST formations. For example, power cars coupled back-to-back were sometimes seen transiting between maintenance locations or on driver training duties and, of course, the inevitable failure of a HST set would sometimes see a diesel locomotive at the head of the formation providing the motive traction for the train.

Secondary Workings

There were many routes that played second fiddle to more prestigious routes in terms of traction and rolling stock, and thus provide interesting subjects for the modeller.

The Liverpool Street to Cambridge and Norwich routes were plied by 9-coach trains running scheduled services hauled by the 'standard' diesel traction of the day (most

Unidentified Class 50 hauls a rake of Mk 2 coaches on the Waterloo-Exeter Service in 1987. (*Jamerail/Flickr*)

commonly a Class 31, 37 or 47 locomotive). This continued until the end of the 1980s, when these services were replaced by multiple units and Class 86 locomotives, hauling Mk 3 coaches operating in push-pull mode with Driving Van Trailers.

Services from London Paddington were briefly hauled by Class 31 locomotives using Mk 1 coaches in the 1970s, before being taken over by Class 47 and Class 50 locomotives utilising Mk 2 stock, these having been displaced from other principal routes by the arrival of the HST.

Waterloo-Exeter, 1987	
Locomotive Number: Class 50	
Vehicle Type	Load
Mk 2A/B TSO	Passengers
MK 2A/B FK	Passengers
MK 2A FO	Passengers
MK 2A/B BSO	Passengers
MK 2A TSO	Passengers
MK 2A BSO	Passengers
MK 2A TSO	Passengers

Further north, the West Highland line produced interesting workings, Class 24 and 25 locomotives gradually giving way to services hauled by Class 37s by the early 1980s. During this period, provincial routes were largely the preserve of the DMU; until the mid-1980s these were often first-generation units. From the mid-1980s, the introduction of second-generation multiple units including the Class 15x and 'Pacer' fleets resulted in the mass withdrawal of first-generation units. However, a number of first-generation units survived until the advent of privatisation and were certainly a defining feature of many routes. One of the most interesting features of multiple units was the ability to shorten or lengthen services to meet demand, something that is rarely depicted on layouts.

An interesting feature of some provincial routes was the temporary substitution of short-formed multiple unit replacement services. During the 1980s, short locomotive hauled formations ran extensively in many areas, primarily as replacements for new 'second generation' vehicles plagued by reliability

A typical DMU replacement service is seen here at Bennerley in 1987. 31433 hauls a rake of five coaches on the Liverpool–Great Yarmouth service. (*Dave Peachey*)

issues. By the late 1980s/early 1990s, reliability issues with these units had been resolved, and the short rakes of coaches hauled by a low-powered single diesel locomotive (usually a Class 31 or Class 33 on the Southern region) began to disappear.

Such services were typically no more than six coaches, in some cases as short as three. There would normally be one brake vehicle and no more than half of one of the coaches would be reserved for first class passengers. Vehicles tended to be a mixture of Mk 1 and Mk 2 coaches (non-air conditioned). A variety of liveries was common, with BR Blue and Intercity liveries being mixed in the same train.

A compact four-coach formation suitable for many layouts would typically be formed thus: BSO–CK–TSO–TSO. Sometimes the CK would be swapped for an FK to provide a small amount of dedicated First-Class accommodation.

Special Passenger Trains

The term 'special' can cover a variety of different trains, from additional trains laid on to cope with the demands of the holiday season, to trains carrying fans to a large sporting event, to charter trains laid on especially for railway enthusiasts, perhaps to commemorate a certain event or celebrate the retirement of a particular locomotive class. Running a special allows us to plausibly operate trains that wouldn't ordinarily be found on that particular line, or in service during that particular era.

The application of a few common-sense rules will enable specials to be 'slotted' into the existing running schedule in a sensible and realistic manner.

- Consider the purpose of the special, this may inform the type of traction and rolling stock to be used in the train. For example, a special service carrying football fans would be unlikely to incorporate significant First-Class facilities but might require additional catering services. Similarly, a service chartered for railway enthusiasts would likely feature some sort of vintage or classic traction, perhaps a preserved steam locomotive.

- Avoid any obvious incompatibilities, for example, air braked coaches running behind a steam locomotive equipped only with vacuum brakes.
- If absolute accuracy is required, particularly when dealing with 'classic' traction, then further research will be required to determine what members of a particular class of steam locomotive were preserved, and whether they were operational and capable of main line running during the desired time period(s). For example, although the famous 'Flying Scotsman' has been in preservation since 1963, it's only actually spent a percentage of that time running on the main line.
- Avoid overuse of specials. Even on a busy main line, it would be unlikely for multiple specials, hauled by different 'celebrity' locomotives to appear all on the same day…. this is a particular weakness of many modellers and many layouts! Remember that the focus should always be on the everyday 'mundane' in order to retain an authentic atmosphere.

Perhaps the first image the comes to mind when we might think of a 'Special'; an immaculate preserved steam locomotive operating on the mainline in the diesel era. In this picture, 5MT No.44932 heads the 'Risborough Venturer' comprising Mk 1 and Mk 1 Pullman stock past Langley Mill on the Erewash Valley line in April 1990. (*Dave Peachey*)

The famous 'Jolly Fisherman' holiday special is an example of a fairly common 'special' and could plausibly be run on any BR Blue or Sectorisation layout set in the East Midlands. The train often featured Mk 1 and early Mk 2 stock and was hauled by Toton based Class 20s to Skegness from a variety of East Midlands Locations. Although the Class 20 was not equipped with Electric Train Heating, the service only ran in the summer months and as such heating was not required. This example is seen passing Shipley Gate in 1991. (*Dave Peachey*)

Railtours are another opportunity to use locomotives not usually found on passenger workings. Here, 58042 hauls the 'Denby Phoenix' Railtour past Breadsall back to St Pancras. The locomotive on the rear is 47976. (*Dave Peachey*)

A 'Special' doesn't even have to feature coaching stock. Here, 31155, with 55015 in tow, leaves Toton with an air-braked freight bound for Derby St. Marys. The 'Deltic' was being moved from the Midland Railway Butterley to the Keighley and Worth Valley Railway. Try running this train at an exhibition! October 1988. *(Dave Peachey)*

Freight Train Formations

Entire volumes can (and have) been devoted to freight operations on Britain's railways, so this book cannot hope to offer more than a flavour for the main types of operation seen during the BR Blue and Sectorisation eras. Inevitably there are workings and details we cannot possibly discuss in such a restricted space.

Although the modernisation plan of 1955 had called for the removal of unfitted freight wagons within ten years, it wasn't until the late 1970s that major inroads were made into the unfitted freight fleet. Unfitted trains lasted until 1985 in revenue earning service, and into the 1990s in departmental service. The major problem with unfitted trains was their low top speed; 25mph for a fully unfitted train. To get around the problem, as many trains as possible were marshalled with a 'fitted head', i.e., a number of vehicles equipped with continuous (most commonly vacuum) brakes which would be marshalled next to the locomotive, with the unfitted vehicles and the brake van at the rear. Such 'partially fitted' trains could run at speeds of between 35 and 45mph.

One of the more unusual inventions of the time was the brake tender. Introduced between 1962 and 1864, brake tenders were designed to increase the available brake force to the driver when an unfitted train was hauled by a diesel locomotive. Essentially consisting of a chassis filled with scrap, and riding on ex-coach bogies, brake tenders would be included at the head of certain heavy workings, primarily coal trains in the Midlands and the Northeast. The last brake tenders seem to have fallen out of use by the mid-1970s and were all scrapped by the 1980s.

The advent of sectorisation in 1985 meant that locomotives and wagons were assigned to different sectors within British Rail. Therefore, it would be extremely unusual (but not impossible) for a coal sector liveried locomotive to be seen hauling a petroleum train. However, many locomotives and most stock were not repainted, and so Blue and early Railfreight liveries continued to feature significantly right up until privatisation (and beyond in some cases).

The Correct Use of Brakevans

No chapter on freight train operations would be complete without a discussion on brakevans as these are a major visual feature of many freight trains and are often incorrectly used on layouts by modellers.

The complicated mix of braking systems in use during the BR Blue and Sectorisation eras, plus changing rules within the railway landscape is something that is quite characteristic of the era. The exact nature of the brake equipment fitted to each vehicle is hugely complicated and is beyond the scope of this book. However, the correct use of a brakevan is achievable for any modeller, and thus this 'rule of thumb' section can be considered a general guide.

Brakevans were provided on freight trains going back to the earliest days of the railway, particularly along goods-only lines. In 1968, a pivotal agreement was reached between the trade unions and BR which permitted guards on fully fitted trains (i.e., a train with all the wagons fitted with either air or vacuum brakes) to ride in the rear cab of the locomotive. This change eliminated the requirement for large numbers of brakevans overnight, but there was still a requirement for a brakevan on unfitted or partially fitted trains.

One area where compliance with this rule could not be observed was when the train was hauled by a locomotive with a single cab, such as a Class 20, for example. In this case a brakevan would be provided, regardless of the composition of the train.

By the end of the 1980s, the widespread adoption of Driver Only Operation had all but eliminated the requirement for guards on freight trains. There were exceptions, mainly trains conveying hazardous goods. Other special cases might also apply, such as particular local moves that required extended propelling of a train, or on routes where there were many manual crossing gates which had to be opened and closed by the guard. This complicated situation lasted until the early 1990s, mostly due to the extended length of time it took BR to standardise on air brakes.

For modelling purposes, it would not be incorrect to stick a brakevan on the rear of a train containing vacuum braked or unfitted wagons (or any mixture thereof). If you are particularly concerned about prototypical accuracy but have no reference material for the train being modelled, then a Bauxite liveried example of the BR 'Standard' 20-ton pattern brake van would be a 'safe bet' for the 1970s through to the early 1990s.

It should also be noted that unfitted and partially fitted engineering trains lasted longer on the network than any other type, and thus if your layout is set in the 1990s, for example, these are really the only type of unfitted or partially fitted train that you should run. There is a wide variety of ex-traffic brake vans in suitable Engineers' liveries available as Ready-To-Run models that can be attached to the rear of such trains.

Coal

For much of the time that the railway has been in existence, the most important commodity moved (apart from passengers) was coal. Until the 1950s, coal provided nearly all the heat, power and lighting for urbanised and rural communities in the UK and nearly all stations received coal deliveries.

From the 1960s onward, including the period covered by this book, coal traffic contracted significantly. However, it remained one of the most important commodities carried by the network and coal trains were an extremely common sight on most areas of the network, in some areas, particularly the Midlands and the Northeast, coal trains still formed the majority of freight workings.

Although the ubiquitous 16 tonne mineral wagon was still in use during this period, the population of such wagons was drastically reduced, particularly the unfitted type which had been completely removed from traffic by

Seen at Heanor in 1980, 20195 and 200006 haul a fairly typical mix of vacuum braked and unfitted coal wagons. Such workings had ceased by 1985, and most of the unfitted and vacuum braked mineral and hopper wagons were scrapped. (*Dave Peachey*)

1985. By the early 1990s, the small remaining number of vacuum braked examples had been transferred to departmental use. The number of unfitted or vacuum braked coal hoppers and other larger types of coal wagons were also reduced, and such workings had ceased by 1985.

The two most common types of coal wagon from the mid-1980s on were the 46 tonne GLW Air Braked 'Merry-Go-Round' (MGR) type hopper and the 45 tonne GLW air braked hopper. Both types could be seen working in large 'block' trains ferrying coal from distribution centres to power stations or stockpiles. They could also be seen working together and in small mixed rakes with other air-braked stock.

From the late 1970s until the mid-1980s, mixed rakes of unfitted or vacuum braked wagons could be seen being hauled by virtually any diesel locomotive. The exact makeup of such trains would often be region specific, as would the locomotive types used to haul them.

Toton – York, March 1975	
Locomotive Number: 25267	
Vehicle Type	Load
MCV 21t Mineral	Coal
HTV 21t Hopper x 5	Coal
HTO 21t Hopper x 25	Coal
CAR Brake Van	-

Over time, it was recognized that a efficient model of unloading coal at power stations was needed, resulting in the MGR system being introduced in 1965. This system utilised locomotives with slow speed control, air braked hoppers and specially equipped unloading equipment at the power station. In theory at least, it allowed for continuous loading and unloading of coal into power station storage bunkers, reducing the handling, time and cost it took to keep the many power stations fed with coal.

A 'Classic' vacuum fitted coal train is seen here at Pye Bridge in 1982. 45040 hauls a train of MDV 21t coal wagons south. (*Dave Peachey*)

HAA hoppers being loaded under the rapid discharge bunker at Bentick Colliery in 1993. The train is destined for Ratcliffe Power Station. (*Dave Peachey*)

After a brief trial, the MGR system was implemented at all large power stations, with only a few locations physically unable to accommodate the new equipment continuing to be served by alternative means.

Gradually, as the old vacuum and unfitted wagons were withdrawn from power station duties, long rakes of MGR air braked hoppers appeared, hauled firstly by Class 47 locomotives, then double-headed Class 20 and Class 37 locomotives were used. The 'new' freight locomotives of the 1970s and 1980s, the Class 56 and Class 58 respectively, were heavily utilised on coal workings upon their introduction, particularly the latter which worked almost exclusively on coal workings until the advent of privatisation (apart from a brief period during the miners' strike of 1985, which, incidentally, saw the withdrawal and scrapping of most of the remaining unfitted and vacuum braked coal wagons). Finally, from their introduction in 1990, Class 60 locomotives were used to take advantage of their tremendous haulage power. Scottish MGR workings were almost exclusively handled by Class 26 locomotives until their withdrawal from traffic in 1993.

Rake sizes varied, with the MGR rakes hauled by Class 47s varying between 30 and 42 wagons depending on region. In order to model a convincing representative train of this type in N Gauge, you would need at least 20 wagons, with a rake of around 30 wagons providing a realistic visual spectacle.

Some trains comprising air braked hopper wagons required a brake van at one or both ends of the train. This was not to provide a guard for the train as per the unfitted or partially fitted trains of yesteryear, but to provide accommodation for groundstaff who were required to operate the manual level crossings prevalent on certain routes, the Denby branch being such a line.

To give an indication of the number of coal trains needed to feed one power station, during the 1970s, the intake of coal at Ratcliffe Power Station was in the order of 20–28 trains *per day*. Therefore, it can safely be concluded that many modellers will find a place for at least one MGR train on their layouts.

58015 Passes Codnor Park with northbound empty HEAs, February 1993. (*Dave Peachey*)

Coal sector liveried 56135 hauls a rake of Russel coal containers, Cramlington, 1991. (*David Ford*)

37897 hauls a rake of PFA Wagons loaded with empty Cawoods coal containers, 1987. (*Jamerail/Flickr*)

Power stations weren't the only users of coal, and the MGR system, whilst designed specifically for use with power station unloading equipment, was of little use in other areas, particularly small sites where the most sophisticated equipment available for unloading might be a few men and shovels, or a digger. This was recognised and British Rail introduced a 32.5 tonne air braked hopper from 1975 onwards (initially coded HBA under TOPS, later HEA). These wagons were very different in appearance to the MGR hoppers and were fitted with manually operated discharge doors.

These hoppers operated in long and short rakes, or sometimes mixed in with other traffic as part of a Speedlink service (see below). Therefore, for those with restricted space, a train comprised of this type of hopper may be more suitable.

By the late 1980s, the traffic that this wagon was built for was in decline, and the wagons were often repurposed into other uses, more specifically as scrap carriers and as barriers for hazardous materials (see elsewhere in this chapter).

Finally, coal on specific flows was also conveyed in specially converted container wagons. There were two main types, the FPA wagon carrying 'Russel' containers which were based on converted chassis reclaimed from VCA Vans and SAA steel carriers, and the short wheelbase PFA purpose-built container wagons used initially by Cawoods Coal Ltd.

1991	
Locomotive Number: 56135	
Vehicle Type	Load
FPA (Russel Container) x28	Empty

Steel

Steel, in the form of strip, coil, bar, girders and ingots was all carried on the network in a variety of general purpose and specialist vehicles. In some areas (such as South Wales and Teeside), steel trains rivalled coal in terms of daily traffic, with entire fleets of locomotives and wagons being assigned to the duty. Many

This bogie strip coil wagon (TOPS code BVW) with plateback bogies, is typical of older vacuum brake steel stock. It was pictured here at Craven Arms in 1988. (Jamerail/Flickr)

different types of wagons have been used for steel traffic over the years, many purpose built specifically for steel traffic.

During the 1970s through to the mid-1980s, many older unfitted or vacuum braked steel carrier wagons were still in use. These were utilised mostly in block trains from steelworks to industrial centres or the docks for export. As the 1980s wore on, many of these services would be withdrawn, mirroring the general drawdown of steelmaking in the UK.

By the later 1980s, older types of wagon had largely been withdrawn or relegated to departmental traffic, and modern air braked stock was used on steel traffic. Purpose built steel carriers, TOPS codes BAA and BBA were primarily used for the transportation of steel coil, with air braked bogie flats (TOPS Code BDA) often being used for ingots, girders and bar. Purpose built steel carrying opens (TOPS Code SPA) were also utilised, often with retractable hoods to prevent corrosion of the newly milled steel during transportation.

Air braked open wagons, such as the OCA, were also utilised on some flows, particularly those associated with export traffic, and could often be seen carrying smaller quantities of steel within Speedlink trains.

Finally, a number of private owner bogie steel carriers were introduced during the 1980s. Built primarily for coil traffic, they featured sliding hoods and air brakes. These wagons would most often be found operating in block trains.

6D19, Lackenby–Wakefield, August 1989	
Locomotive Number: 37510	
Vehicle Type	Load
BBA Bogie Steel x5	Steel Coil
KIB Bogie Steel Hood x5	Steel Coil

Petroleum

With increased car ownership and extensive use of plastics in the 1960s, demand for petroleum and petroleum-based products increased dramatically. British Rail signed haulage contracts with all the major oil companies, and the tonnage in petroleum products rose to over 15 million tonnes per annum by 1969.

In Trainload Metals livery, 37516 is seen here hauling the Etruria to Tees Yard 'Girder Train' past Breadsall in August 1991. The wagons are air braked bogie flats which were often used for this type of traffic. (*Dave Peachey*)

With this rapid increase in business came a promise to invest in suitable rolling stock. Obsolete tank wagons were replaced by 45 tonne twin axle and 100 tonne bogie tank wagons, which allowed higher operating speeds and were easier to maintain. With this fleet expansion came investment in depots and operating sites, and by the 1970s a network of trains linked together in excess of 300 inland distribution sites with over 30 refineries.

Shell and British Petroleum had around a 50 per cent market share, and by 1969 had a fleet of around 5,000 wagons. Esso was another major player, with around a 25 per cent market share. Therefore, a petroleum train operating on a layout would more often than not carry the logos of one or more of these companies. It should be noted that the wagons were owned by their respective companies and would not normally be seen in mixed workings with one another. The exception to this was that Shell and BP operated a joint fleet of tank wagons until 1976, when the fleet was split 60/40 in Shell's favour. After 1976, each company applied their own logos and branding to their wagons.

In addition, Esso often hired tank wagons from other companies including Algeco, BRTE and Tank Rentals, so it was not uncommon to see an Esso train comprising tank wagons coming from different companies.

Most classes of diesel and electric locomotives have been used on petroleum services over the years, however, the Class 47 was ubiquitous across the network and was the locomotive type most often seen on Petroleum trains until the introduction into service of the Class 60 in 1990.

Other classes commonly used on petroleum trains were Class 37s operating trains from London, Class 33s from the Isle of Grain and Class 25 and Class 31 locomotives from Stanlow and Immingham respectively.

In Scotland, smaller locomotives were usually used, commonly Classes 20, 26 and 27. From the mid-1980s onwards, a number of locomotives from the Class 31, 37, 47 and 60 fleets were allocated to the petroleum sector and decorated with the relevant branding.

Although petroleum trains were more usually comprised of similar wagons, they may not have all been carrying the same commodity. Some clue as to the content of each wagon in a rake can be gleaned from the HAZCHEM panel, but, right up until the 1990s, most petroleum wagons in a train would display just the UN substance number '1270' on a panel, which was merely a generic number covering a number of products.

A Class 40 hauls 6E37, comprising Class A two axle and bogie tanks, past Castleford in 1976. (*David Ford*)

In broad terms, there were two classes of petroleum product carried by rail; Class A – lighter petroleum fractions obtained during the refining process. Characterised by high flammability, with a flash point below 23°C, these included naptha, motor spirit, aviation fuel and white spirit; and Class B – inflammable liquids with a flashpoint between 23°C and 60°C. These included kerosene, bitumen and diesel.

Until 1990, all Class A tank wagons had a light grey barrel and red solebar, making identification relatively simple. A change in the rules meant that companies were free to incorporate colourful liveries, such as the yellow and green livery of BP. Class B wagons were normally painted black.

Class A trains often had barrier wagons between the tank wagons and the locomotive and guards vans, to help protect staff in the event of a fire.

6S47, Fawley–Polmadie, March 1972	
Locomotive Number: 47278	
Vehicle Type	Load
TEA 100t Bogie Tank x8	Empty

7F26, Daiston–Stanlow, April 1989	
Locomotive Number: 47278	
Vehicle Type	Load
TEA 100t Bogie Tank x8	Empty

26001 Approaches Millerhill with a short train of Class A tank wagons, 1983. (*Jamerail/Flickr*)

TRAIN FORMATIONS • 45

31466 with a short rake of Class B tanks, carrying Bitumen, seen here at Fairwood Junction in 1987. (*Jamerail/Flickr*)

Welbeck Colliery Junction 1990, an unidentified petroleum sector Class 47 passes with a rake of Class A tanks. (*Dave Peachey*)

47010 hauls a rake of bogie and two-axle tank wagons destined for Port Clarence past Langley Mill on the Erewash Valley Line in 1991. Note the extremely dirty state of the wagons which was fairly typical, a lack of cleaning and spillage of payload contributing to paintwork deterioration. (*Dave Peachey*)

Cement

Cement trains have been a feature of the network for many years. Cement trains normally operated as 'block trains' (i.e., a train carrying one commodity); although they could be seen in mixed rakes, this was less common after the mid-1970s. Until the late 1970s and early 1980s, the most common types of wagon seen in cement trains were the PCV 'Cemflo' and the CPV 'Presflo' types, both vacuum braked. There was also a more rarely seen wagon called a 'Prestwin' (TOPS Code CQV). The 'Cemflo' type in particular were problematic, as the suspension was degraded by the cement dust, causing accidents. The Cemflo wagons were withdrawn from service in 1988, with the Presflos being withdrawn in 1987. The Prestwin wagons were less successful and lasted in service until 1983. Another common type was the 'Chevron' or depressed centre tank, the first of which were introduced into traffic in 1966 and were operated on behalf of the large cement producers such as Blue Circle, Ketton and Rugby, often sporting distinctive owner liveries.

Warrington Bank Quay, 1983	
Locomotive Number: 37290	
Vehicle Type	Load
PCV Cemflo x26	Cement

In the early 1980s, private owner cement wagons began to be introduced to traffic, principally those carrying the TOPS code PCA, which covered a number of privately built types including 'straight' and 'depressed' centre types. These wagons, along with the 'BR' PCAs could be seen running in mixed rakes well into the 1990s.

37290 Hauls a rake of Cemflo cement wagons past Bank Quay Station, Warrington, 1983. (*Jamerail/Flickr*)

Unidentified Class 33 hauls a rake of Rugby Cement liveried PCA tanks past Worthing Junction in 1987. (*Jamerail/Flickr*)

47325 hauls a short rake of PCA wagons between Sheet Stores and Stenson Jn, 1991. (*Dave Peachey*)

Worthing Junction, 1987	
Locomotive Number: Class 33	
Vehicle Type	**Load**
PCA x18	Cement

In addition to bulk powdered cement, a considerable quantity of bagged cement traffic was transported in covered vans until the late 1970s. APCM bagged cement traffic, with its distinctive long-wheelbase vacuum-braked vans had finished by 1980, with the wagons stored and either modified for other purposes or scrapped by 1987. A small amount of bagged cement was also transported in newer air braked vans during the Speedlink era.

Aggregates and Other Mineral Traffic

Construction projects use large quantities of aggregates, and various flows could be seen in all areas of the country. The traffic includes a variety of heavy minerals and can include anything that is mined or quarried (other than coal). The various traffic flows varied tremendously, and it is impossible to cover it in depth within this chapter.

Aggregates' growth in consumption starting in the 1960s mirrored the resurgence of bulk rail freight. The demand for aggregates was in part fuelled by the construction boom associated with road development and the investment in public and private building projects, particularly in the south. Demand for construction materials quickly outstripped local resources, and quarries in the Mendips in Somerset provided much of the material for construction projects in and around London.

Prior to the introduction of the Class 56 and later the Class 58, 59 and 60 locomotives, traditional vacuum-braked tippler and hopper wagons were used in shorter block formations hauled by Class 33, 37, 40, 45 and 47 locomotives. These types of workings, although increasingly supplanted by longer Type 5 hauled air braked block trains, continued to be seen into the late 1980s.

The arrival of the Class 59 locomotives on Mendip workings in the mid-1980s revolutionised these particular traffic flows. Paired with modern air braked tippler wagons, these locomotives were capable of hauling huge loads with unheard of reliability, and some of the heaviest freight trains on the network could be seen on these workings.

With the wind down of the steel industry, the demand for iron ore gradually decreased and by the 1980s had largely disappeared. A few workings carried on which can provide an interesting source for modellers, but these were geographically specific, such as the workings to and from Redmire. In contrast to aggregates, iron ore workings tended to occur in shorter

45070 is seen here hauling a rake of HJV hoppers past Duffield in 1987. The hoppers are loaded with sugar stone. The appearance of Class 45s on mineral trains was rare by this point and 45070 was withdrawn and scrapped less than 12 months after this photograph was taken. (*Dave Peachey*)

56048 heads past Whitwell with a rake of tippler wagons carrying quarried stone in July 1991. The train is headed for Worksop. (*Dave Peachey*)

train formations. When the freight business was sectorised in the early 1980s, iron ore could be operated by more than one 'division' in contrast to workings such as coal, obviously only operated by the coal sector.

Principally, iron ore traffic featured short rakes of ore hoppers hauled by diesel locomotives, most commonly a Class 37. Other industrial type minerals including salt, lime, sand, gypsum and soda ash were commonly transhipped by

An unidentified Class 59 (probably 59003) hauls a rake of Yeoman air braked tippler wagons past Fairwood Junction in 1987. (*Jamerail/Flickr*)

37051 Hauls a rate of tippler wagons loaded with dolofines (quick lime) past Buston Barns in 1987. (*David Ford*)

rail. Many of these substances have a form akin to powders and could be discharged or loaded using pressure in a similar manner to cement. Therefore, wagons designed for the carriage of cement such as the Prestflo, and Prestwin types often featured on such workings.

Grain

Perhaps the defining grain wagon of the mid BR era was the fleet of private owner air-braked grain hoppers owned by the leasing firm BRT. These wagons, first introduced in 1965, gradually replaced existing BR grain wagons on Cross-Border grain traffic. The BRT wagons, later coded PAF under TOPS, with their blue livery and large advertisement boards for the various distillers, raised the profile of grain traffic considerably. A smaller fleet, introduced on behalf of Associated British Maltsters, was painted yellow and was adorned with similar boards.

Trains comprising the BRT hoppers were largely block flows, centred around a number of loading points in the Eastern Counties and a corridor on the East Coast Main Line. The wagons were worked from various locations and formed into block trains at Doncaster, before their forward journey to Scotland, where they would typically be split again into smaller rakes to be worked out to the various distilleries. Elsewhere, smaller grain flows continued to use ageing BR 20t grain hoppers well into the 1980s.

By the mid-1970s, Anglo-Scottish grain traffic was in decline, and the BRT fleet was experiencing significant under-utilisation. However, the market for grain traffic was still considered attractive and in 1980 BR launched the 'Grainflow' initiative. Grainflow was essentially a partnership between BR and the wagon owner TSL, who already had experience of bulk grain traffic in Europe. The primary objective of Grainflow was to bring existing and future grain traffic flows under the control of the Speedlink network using high-capacity air-braked wagons. The design chosen was the Polybulk type hopper. This wagon, in the familiar green and grey Grainflow livery, quickly became a feature of cross-border Speedlink services upon their introduction in 1974. In 1983, a further batch of Polybulk hoppers were constructed, including some to a modified design for Distillers, operating in their grey and blue livery. The arrival of this second batch of Polybulks finally finished off the remaining BRT hoppers, which were cascaded to other types of traffic (such as sand) before eventual withdrawal.

Elgin, September 1980	
Locomotive Number: 47322	
Vehicle Type	Load
PAF Grain Hopper x32	Empty

Chemicals

Chemicals traffic really took off in the 1960s, as major manufacturers such as ICI and BP introduced fleets of specialised tank wagons, some of which were painted in bright liveries, and which ran in block trains between key sites. This investment continued into the 1980s, with some companies applying for government grants to provide new fleets of wagons and improved facilities.

The Speedlink network allowed for small scale traffic flows, some of which eventually expanded and became flows requiring block trains. The ending of the Speedlink service in 1991 deprived the chemical industry of flexibility, and left customers with a choice, either pay for a full block-train service, or use the road network. Unsurprisingly, smaller flows all moved to the road network and by the mid-1990s, most chemical traffic had disappeared from the network.

One of the largest centres for chemical traffic during this period was Teeside. Haverton Hill Yard, on the north bank of the River Tees, served the Billingham Chemical plant which was operated by ICI. Block trains of anhydrous ammonia ran to Barton-on-Humber, Severn Beach, Heysham, Leith and Grangemouth. The trains consisted of white or grey bogie tank wagons sporting a horizontal orange band. These wagons carried a variety of TOPS codes including TCA, TDA and TIB.

Haverton Hill also produced a variety of other flows, including methanol, carbon dioxide, urea and amines. Another plant at Seal Sands produced a variety of flows, with a regular service to the UK Fertiliser plant at Ince and Elton, and an inward train of sulphuric acid which originated at St Helens.

On the south bank of the Tees, the ICI Wilton plant produced cyclohexane which was dispatched daily to Stevenson, as well as regular trains carrying adipic and terephthalic acids. Further south, there were a number of rail served plants around the Humber Estuary, including BP Chemicals at Saltend, which produced acetic acid which was transhipped to Baglan Bay, and Celanese near Derby.

In the Northwest, there were a number of plants associated with the industry including ICI at Northwich, BP at Sandbach, ICI at Runcorn, UKF at Ince and Elton and Shell and Associated Octel at Ellesmere Port.

ICI Northwich produced quantities of soda ash, also known as sodium carbonate, for use in glass and detergent manufacture, and a major portion of this product was moved by rail. Until the late 1970s, the majority of the soda ash was moved in covered hopper wagons (TOPS codes CHO, CHP, CHV) via the standard wagonload network, but by the late 1970s, ICI had leased a large number of air-braked private owner wagons, such as pressure discharge tank wagons (TOPS code TTA) and greater use was made of dedicated block trains.

The BP works at Sandbach was dispatching chlorine to Esso at Fawley and BP at Baglan Bay until the 1980s. The works also produced caustic soda, hydrochloric acid and sodium hypochlorite.

The ICI Runcorn works was closely associated with the freight-only Folly Lane branch (which was actually electrified purely to service ICI traffic). Runcorn produced mainly caustic soda

40076 heads 6P42 past Wigan Northwestern, a train of tank wagons headed for the ICI works at Corkicle in 1978. The wagons are likely carrying Soda Ash from Northwich. (*David Flitcroft*)

37113 heads a train of acetic acid (vinegar) tanks past Bennerley in 1983. This train was noted for its distinctive smell! The tanks are headed for British Celanese Spondon, near Derby. Also, note the rapid discharge coal bunker under construction at the Shilo Disposal Point on the right. (*Jamerail/Flickr*)

but also a number of chlorinated hydrocarbons such as perchloroethylene and sodium hypochlorite. Block trains were operated from Runcorn to Holywell Junction, Burn Naze, Corkickle, Immingham, Seal Sands, Stevenson and Willesden until the mid-1980s.

The UKF fertiliser plant at Ince and Elton acquired its famous air braked fertiliser fans in 1968. By the mid-1970s, a fleet of over ninety vans were in service, and scheduled trains were run to a number of locations.

As the 1980s progressed, changes in production and distribution patterns, the introduction of pipelines and of course increased competition from the road network put increased pressure on the viability of chemical traffic. Norsk Hydro, introducing block fertiliser trains between Avonmouth and Leith, using air braked ferry vans of continental designs, were one exception in a period marked by decline.

Some traffic survived the end of Speedlink in 1991, and acetic acid traffic from Saltend survived until privatisation.

7D05 Amlwch to Llandudno Junction, 1985	
Locomotive Number: 47128	
Vehicle Type	Load
OBA Open	Barrier Wagon
TTA Tank x4	Ethylene Dibromide
TTB Tank x4	Liquid Chlorine
RBX ex-Ferryvan	Barrier Wagon
CAR	Brake Van

Automotive

Block workings of new cars were often conveyed on Carflats, many of which were converted from the underframes of redundant coaching stock. A dedicated fleet of carflats was owned by the leasing company Railease and used well into the 1980s.

In addition, articulated 'Cartic-4' wagons were introduced from 1964, which, as the name suggests, were semi-permanently coupled

4-wagon sets with articulated bogies and, unusually for the time, fitted with air brakes from new. These wagons offered a step change in capability, with the ability to transport twice the number of cars as the 'carflats'. The original BR wagons were followed a few years later by a fleet of privately owned wagons to a very similar design, owned by the distributors MAT, Toleman and Silcock.

From 1968 onwards, there were some flows of parts between various manufacturing centres, such as engines between Longbridge and Cowley, or body panels between Swindon and Longbridge. These components were normally carried in long wheelbase air braked vans. Finished cars for export were loaded at several distribution sites around Birmingham and destinations included the ports at Dover and Harwich.

BR carried components for Ford between Halewood and Dagenham, using a fleet of dedicated air braked vans. This route often featured electric traction between Halewood and Willesden.

At its peak in 1972, the British Car Industry produced 1.9 million vehicles annually, but the rest of the decade saw the industry dogged by industrial action and poor productivity, with a corresponding drop in automotive traffic on the rail network. As domestic car production fell, the import of vehicles started to increase, from 7 per cent in 1970 to 34 per cent in 1980. The railway supported this trend by setting up flows from various ports, often supported by the Speedlink network for flows not requiring the services of a block train. Some examples of these flows were imported Datsuns from Eastleigh to Leyburn, and various Japanese cars from Lowestoft.

Newer stock was introduced in the 1980s, the 'Autic' (TOPS Code PKA) was introduced in 1981 and improved efficiency by including an integrated ramp to speed up loading operations. A single deck version of the Autic called the Comtic was also introduced in 1981 and was used to transport commercial vehicles such as vans and chassis cabs which were too high for the standard double-deck car carriers. These wagons could be found on a variety of automotive traffic flows.

Peasecliffe, July 1983	
Locomotive Number: Class 31	
Vehicle Type	Load
XMA Cartic-4 x2 (8 wagons)	Empty

47346 seen here with a train of empty Cartics passing Hornsey Carriage Sidings in 1981. (*Jamerail/Flickr*)

This unidentified Class 31 Locomotive has just left Peasecliffe Tunnel with a rake of empty Cartics. July 1983. (*David Ford*)

37266 is about to depart Leith in 1987, with a train load of new Fiat cars. Note the screens on the sides of the Cartic wagons. These were designed to reduce the potential for damage to be caused to new vehicles, both accidental (flying ballast and tree branches) and deliberate (vandalism). (*David Ford*)

Commercial Explosives

Throughout the 1970s, boxes of explosives were carried in BR 11 ton Gunpowder Vans (TOPS Codes CXO/CXP/CXV). Most of the traffic originated at the Nobel factories in Annan, Ardeer, Gathurst, Pnrhyndeudraeth and Snodgrass and from Explosives & Chemical Products Limited at Alfreton.

A trip working ran between the works at Ardeer and Snodgrass, usually hauled by a shunting locomotive. This working would typically comprise nine or ten gunpowder vans, but the majority of the traffic was forwarded in single vanloads to various destinations around the country, more specifically to serve the numerous collieries and quarries which required large quantities of commercial explosives for blasting activities.

Among depots frequently served were Ashford, Bogside, Buxton South, Callerton, Crewe, From, Llantrisant, Okehampton, Redhill, Stainforth & Hatfield, Tonbridge, Truro and Wednesbury. There were also movements to Aberdare, Cardiff, Morpeth, and Walsall.

The number of destinations for explosives trains gradually dwindled during the 1980s, and by the end of 1989 only Drinnick Mill in Cornwall still saw deliveries of commercial explosives by rail. Inter works moves between Annan and Snodgrass had ceased by the start of 1991.

The traditional vacuum-braked gunpowder vans – with some vehicles dating back to pre-nationalisation – were incompatible with the British Rail Air Braked Network/Speedlink and therefore modernisation was required. From the late 1970s onwards, long-wheelbase air-braked vans began appearing on explosives traffic, before taking it over completely in 1983, with a special pool of forty such vans being formed.

Only wagons that were controlled by the Central Wagon Authority were permitted to carry explosives. Several types could be seen carrying military munitions and stores, but the only type of wagon authorised for the carriage of commercial explosives were the aforementioned long-wheelbase air braked vans (TOPS Codes VAA, VBA). Vehicles in explosives traffic underwent more regular checks than standard wagons, and due to this, many vans fitted with experimental suspension were utilised so that regular checks of the suspension could be made.

25152 seen here at Langley on the Erewash Valley Line in 1982. The vans are being worked from Chemical Products at Alfreton. It is likely that the rear van contains explosives, the two ex-Ferry vans are probably barriers. (*Dave Peachey*)

Barrier wagons were required to separate loaded vans from the locomotive and the rear of the train, several ex-Ferryvans (coded RBX) were converted and could be seen on workings in the 1980s. However, it was not unheard of for other spare wagons to be utilised, in theory any wagon, as long as it was empty, could be utilised as a barrier vehicle.

7G35 Pwllehi – Bescot, September 1979	
Locomotive Number: 25324	
Vehicle Type	Load
ZXP Flatrol	Empty
QPV Staff and Dormitory Coach	Empty
ZQV Tool Van	Departmental
VVV 12t Van x5	Empty/Barrier
CXV Gunpowder Van	Commercial Explosives
RBV (ex-Banana Van) x2	Barrier
CAO 20t Brake Van	-

7F16 Llandudno Junction – Warrington, February 1988	
Locomotive Number: 47214	
Vehicle Type	Load
SPA Steel Open	Sawn Timber
PIB Cargowaggon Van	Aluminium Ingots
TAT Class B Tank	Empty
TTA Class A Tank x 2	Empty
HEA Hopper x3	Empty
VAA Van x2	Commercial Explosives
RRA Barrier (Ex-SAA)	Barrier

The Air Braked Network (ABN) and Speedlink

The first services utilising air-braked stock, marketed as the 'Air Braked Network' (ABN) were introduced between Bristol and Glasgow in 1972. ABN was rebranded to Speedlink in 1977.

By 1985, when the remaining traditional unbraked and vacuum-braked goods trains were being withdrawn, a Speedlink network of almost 150 trains had been established. The network linked approximately a dozen or so large yards and around twenty smaller terminals. Trains were not marshalled in the traditional sense; a train would stop at an intermediate yard only to exchange pre-formed sections of a train so that, in theory, a wagon would only be shunted twice, once in its originating location and once again when it reached its destination.

The three busiest Speedlink routes were the East and West Coast Main Lines and routes between the Northeast and the Midlands. Consists were varied, and could contain gas, oil, scrap, cars, steel, perishable and fragile goods, grain, and cars. Trains could be lengthy, or as short as a single wagon depending on network needs.

Initially, the Speedlink fleet consisted mainly of British Rail owned air braked stock, but by the early 1980s an increasing proportion of wagons in Speedlink trains comprised private owner wagons. As many private wagons were loaded and unloaded at private sidings, there was also an increase in 'trip' workings whereby loaded and empty wagons would be moved between the private owner site and the nearest Speedlink yard.

In addition to regular freight, the Ministry of Defence (MOD) made extensive use of the Speedlink Network. Indeed, there were even dedicated Speedlink services purely to handle military traffic.

7E49 Bescot – York, May 1985	
Locomotive Number: 37066 and 37065	
Vehicle Type	Load
TTA x3	Traction Gas Oil
POA x5	Steel Scrap
BDW x2	Steel Section
BDA	Empty
YGH	Empty
OBA x4	Steel Tubes
PGA	Empty
ZRA	CM&EE Stores
POA x9	Steel Scrap

37232 is seen here at the Gloucester Loop with a Speedlink Service comprising VDA air braked vans. (*Jamerail/Flickr*)

6B68 Inverness – Mossend – Millerhill, July 1989	
Locomotive Number: 37261	
Vehicle Type	Load
OCA Open	Empty
PCA Tank x4	Empty
FPA x4	Empty
PCA Tank x3	Empty
OTA x4	Timber

Parcels and Newspapers

In 1977, all remaining pre-nationalisation stock was withdrawn, leaving the bulk of the parcel and newspaper services to be operated by Mk 1 coaches or Mk 1 GUVs. There was also a number of specialist parcels DMUs, some purpose built, and some converted cheaply from redundant passenger vehicles. The Class 128 was probably the most well-known unit, lasting in service until the late 1980s.

By the mid-1980s, the parcels business was facing competition from road transport. Underfunded and blighted with outdated and inefficient procedures, the rolling stock utilised by the parcels business was elderly and unreliable. In 1986, Royal Mail announced they would be ending their contract to carry parcels with BR, and, to compound matters further, between 1986 and 1988, newspaper traffic was also lost. The net result of this loss in traffic was a mass withdrawal of all newspaper and a large majority of parcels stock.

Throughout the 1970s and 1980s, newspaper and parcels traffic operated primarily between principal stations, with some of the larger stations having dedicated facilities including platforms or bays dedicated to the loading and unloading of parcels services.

If you are modelling a main line station during this era, then incorporating a parcels platform can add some operational interest to your model.

Boreham Up Parcels 1975. The train is comprised entirely of Mk 1 CCTs, GUVs and BGs. (*David Ford*)

Swayfield Down Parcels January 1976. Note the number of pre-Nationalisation vehicles in the train. (*David Ford*)

A fine example of a late BR era parcels train, Network SouthEast liveried 47530 hauls a rake of variously liveried Mk 1 BGs and GUVs past Breadsall on the Newcastle-Plymouth parcels service in 1991. Note the difference in roof profiles between the BGs and GUVs. (*Dave Peachey*)

4H15 Shrewsbury Abbey Foregate–Manchester Mayfield June 1976	
Locomotive Number: 40133	
Vehicle Type	Load
LMS BG	Parcels
Mk 1 GUV	Parcels
Hawksworth BG	Parcels
BR CCT	Parcels
GWR Fruit D x3	Parcels

Hemel Hempstead, September 1979	
Locomotive Number: 86206	
Vehicle Type	Load
Mk 1 BG	Newspapers
Mk 1 BG x3	Parcels

Newcastle – Plymouth, 1991	
Locomotive Number: 47530	
Vehicle Type	Load
Mk 1 GUV x2	Parcels
Mk 1 BG	Parcels
Mk 1 GUV x2	Parcels
Mk 1 BG x4	Parcels

Mail and The Travelling Post Office (TPO)

The Travelling Post Office (TPO), a particular rail service for the sorting of mail en route, had its origins in the first ever postal movement performed by the Liverpool and Manchester Railway in 1830. The Railways (Conveyance of Mails) Act of 1838 made it a legal obligation for the railways to carry mail, and in order to maximise efficiency, specialised rolling stock was quickly developed with onboard facilities such as racks, shelving and desks for the sorting of mail.

Over the years, the TPO concept was refined and improved, with innovations such as the development of lineside apparatus for picking up and dropping off mailbags whilst moving; and subsequently, use of entire trains dedicated solely to mail traffic. By 1926, there were over 100 TPO services operating in the UK,

By the 1970s, however, mail traffic was in steep decline. The use of lineside TPO apparatus was discontinued in 1971, and by the 1980s mail traffic was at an all-time low for the network. In order to improve efficiency and revitalize the traffic, BR formed Rail Express Systems (RES) in 1982. Ostensibly, RES was to rejuvenate the market by streamlining and centralizing mail services, and TPO coverage was reduced to key routes by the time the 1990s had arrived. The final project undertaken by RES was the introduction of Class 325 Dual Voltage EMUs in 1995, to replace Mk 1 based stock, shortly afterwards RES became the first part of BR to be privatised.

Mail continued to be conveyed outside of the TPO system throughout the BR and Sectorisation era, albeit in increasingly smaller amounts. Locomotive hauled services hauled

A rare visitor to the Erewash Valley Line, 47523 seen here hauling a TPO set near Langley in 1991. The train was 5Z36, hauling the TPO to Worksop for display at an open day. (*Dave Peachey*)

86208 is seen here at Eaton Lane with the King Cross–Edinburgh Mail in 1996. (*Dave Peachey*)

dedicated non-TPO mail trains, and smaller amounts of mail were often conveyed in the luggage compartments of multiple units.

Ultimately, the story of mail in the BR and Sectorisation eras is one of managed decline, and dedicated TPO services became increasingly rare at non-mainline locations as the years progressed.

In most cases, TPOs operated with normal Mk 1 Brake (BG) and General Utility Vans (GUV) in their formations to provide additional stowage for mail bags. In 1986 the first Royal Mail 'red' liveries were applied to the stock, although for a number of years mixed formations of blue-grey and red stock were common.

London Paddington, 1988	
Locomotive Number: 50030 *Repulse*	
Vehicle Type	Load
NJX	Mail
NTX	Mail
NUV	Mail
NSX x3	Mail
NUV	Mail
NJX	Mail

Departmental and Service Trains

The railway used an array of different vehicles to maintain the network, carrying everything from new rails and sleepers to new locomotive batteries. Each of the railway departments had a requirement for such traffic, but the most common workings were those on behalf of the Civil Engineers. Trains carrying new ballast, spoil, rails, and sleepers were very common across the entire railway network. There was also a small, but important network called 'ENPARTS' which transported rolling stock parts and components between main works and maintenance depots, primarily in the Western Region. There were also specialist workings such as trains conveying inspection saloons for route examinations or the trips for senior management.

Traditionally, departmental traffic had been carried on the old wagonload network, but the decline of the traditional stopping freight and closure of small goods yards led to the introduction of dedicated departmental services. A handful of vacuum-braked departmental trains branded 'General Utility Services' continued until 1985. This service

Inspection Saloons make for interesting short trains and can reasonably be justified on just about any layout. Here, 73105 hauls an Inspection Saloon through Clapham Junction on 27 July 1988. (*David Ford*)

was mostly used to convey stock not fitted with airbrakes to and from repair or to scrapyards for disposal.

Departmental trains have the distinction of being the last hold-out of unfitted wagons, with partially fitted departmental trains operating until at least 1992. By the mid-1990s, the entire engineers' fleet was fitted, mostly with air-braked stock.

The major bonus of departmental train is that, with one or two exceptions, they can be justified on just about any layout, from the smallest branch terminus to a representation of the East Coast Mainline. Their variety, and the fact they were some of the last unfitted trains on the network, make them especially attractive for enthusiasts.

Toton, 1978	
Locomotive Number: Class 40	
Vehicle Type	Load
ZLV Herring	Ballast
YGH Sealion	Ballast
ZFV Dogfish x7	Ballast
ZLV Herring	Ballast
ZFV Dogfish x8	Ballast
ZEV Catfish	Ballast

July 1983	
Locomotive Number: 37061	
Vehicle Type	Load
ZTV 20t Brake Van	-
YBQ Sturgeon x4	Track Panels
ZTV 20t Brake Van	-

April 1987, Pye Bridge	
Locomotive Number: 31466	
Vehicle Type	Load
ZEV Catfish	Empty
ZFV Dogfish	Empty
Grampus x2	Empty
ZJV Mermaid	Empty
ZFV Dogfish	Empty
ZEV Catfish	Empty
ZJV Mermaid	Empty
ZEV Catfish	Empty
Grampus	Empty
Mullet	Empty
VBA Van	Empty
ZFV Catfish x2	Empty
ZHV (ex 16t Mineral) x8	Empty
ZTO 20t Brake Van	-

37061 hauls a train of Sturgeon wagons carrying track panels through Doncaster Station, 1983. (*Jamerail/Flickr*)

Ballast Trains could be quite short. Here, 47633 hauls a rake of ballast hoppers past Edinburgh Princes Street Gardens in August 1987. (*David Ford*)

31466 passes Pye Bridge on the Erewash Valley Line in 1987, hauling a mix of departmental wagons. The fresh paint on many of the wagons means they only recently left the works, probably at Toton. (*Dave Peachey*)

Stenson Junction, 1994. 37095 hauls a rake of Mermaid side-tipping ballast wagons. (*Dave Peachey*)

Intermodal

Container traffic as we would recognise it came about because of the introduction of the 'Freightliner' concept to the network in the mid-1960s. BR developed a system of transported standardised ISO containers between terminals at high speed in 1963, and between 1964 and 1971 a fleet of dedicated Freightliner wagons (TOPS codes FGA and FFA) were introduced. The wagons were operated in semi-permanent formations with only the outer wagons equipped with buffers and drawgear. The wagons were air braked with a usable payload platform length of 42ft 6in.

By 1985, the Freightliner system had reached its zenith, and there were a large number of Freightliner terminals at ports and other strategic locations.

Freightliner Terminals in 1985		
Aberdeen	Barton Dock	Dagenham
Dundee	Garston	Tilbury
Coatbridge	Seaforth	Harwich
Glasgow	Aintree	Ipswich
Greenock	Holyhead	Felixstowe North
Edinburgh	Nottingham (Beeston)	Felixstowe South
Newcastle	Birmingham Landor St	Cardiff
Stockton	Dudley	Swansea
Leeds	Willesden	Bristol
Hull	King's Cross	Southampton (Mill Brook)
Longsight	Stratford	Southampton Maritime
Trafford Park	Barking Ripple Lane	

The traffic continued to remain relevant and important into the 1990s and in the run-up to privatisation, with a number of newer private-owner Intermodal wagons being introduced throughout the 1980s and early 1990s. These could be seen often on specific traffics, including containerised variants of existing traffic such as gypsum.

58038 hauls a rake of KFA intermodal wagons carrying gypsum containers on the Bennerley Down Fast in 1994. (*Dave Peachey*)

Other Traffic

Ministry of Defence (MOD)

Military trains have been a feature of UK railways for many years, and featured heavily on certain routes. Until the late 1970s, military vehicles were transported on specially constructed wagons called 'Warflats' and 'Warwells' respectively. Equipped with diamond bogies and fitted with vacuum brakes, these wagons dated from the Second World War or earlier.

The increasing maintenance and operational burdens associated with such an ageing wagon fleet resulted in the refurbishment of the majority of the Warwell fleet in the late 1970s, and fitting air brakes and new bogies to modernise the wagons. A small number of unmodified wagons were kept for internal use on military bases.

A new batch of Warflat wagons was constructed by BR Shildon between 1976 and 1981. These wagons were dual braked (TOPS Code PFB), and subsequently converted to air brake only in the 1990s.

To load and unload vehicles to and from Warflats and Warwells, end loading via a special facility was required. However, for locations where such a feature was not feasible, a special type of ramp wagon was required, which could hinge at one end and 'kneel' allowing vehicles to be driven up the wagon and onto an adjacent wagon. This wagon would most commonly be seen in idle in sidings in or near to MOD facilities.

MOD traffic fell under the Speedlink network until its demise in 1991 and was primarily conveyed by trunk services between main yards, where a train would then be broken up and trip worked to its final destination (most often an MOD facility). The majority of

A mixed MOD train seen here at Worthing in 1987. The train is carrying Saxon APCs on Warwells and Viking Tracked Vehicles on Warflats. The inclusion of a van into such a train was common, and such a wagon often carried slings, chocks etc or military muntions. (*Jamerail/Flickr*)

active rail-served MOD bases were located in Southern England and Scotland by the 1980s, so it is unsurprising that MOD trains were most commonly seen around local yards including Eastleigh, Didcot, Severn Tunnel junction, Gloucester, Bescot, Basford Hall, Kingmoor and Mossend.

Trunk services between yards provide considerable variety for the modeller. Trip workings could be short, and it was not uncommon for the locomotive to be longer than the train.

From the 1990s onward, intermodal wagons were sometimes used on MOD workings, with guards provided for any particularly sensitive loads. To facilitate this, two ex-Mk 1 bullion coaches were converted into Escort vehicles, one of which was painted in a green livery with large Railfreight Distribution logos.

Bescot–Donnington, 1985	
Locomotive Number: 20164 and 20131	
Vehicle Type	**Load**
OCA	Empty
VDA	Munitions
VEA	Munitions
PFB Warwell x5	Saracen APCs

Snowploughs

Snowploughs, by their very nature, spend the majority of their lives idle, only being used in periods of prolonged and heavy snowfall. Thus, in the UK, snowploughs have very long lives. Pre-nationalisation designs were largely obsolete and by the 1960s it was clear that a new design of plough was needed for service on British Railways. The BR Independent Snowplough was the result. Built by three different BR works, these ploughs were designed to deal with heavy drifts and were built on the frames of withdrawn steam locomotive tenders. The ploughs were initially fitted with vacuum brakes (later upgraded to air) and also featured a screw type handbrake.

The ploughs were permitted to be marshalled to those locomotives which were normally permitted to use the section of the line concerned, with the exception of Class 40, 44, 45 and 46 Locomotives, which were unsuitable for use with ploughs.

A drawbar was provided for emergency use for example, parking the plough in sidings, but in normal service the plough was coupled to the front or rear of a locomotive and secured by the screw coupling provided.

40172 shunting the earlier type MJ flask wagons at Sellafield in 1975. Note the barrier vehicles between the wagons and the locomotive, in this case spare shock wagon and ventilated van. There would have also been barrier wagons between the wagons and any attendant brake van. (*John Ford*)

7Z00 Winfrith–Gloucester, September 1986	
Locomotive Number: 33028	
Vehicle Type	Load
RBX	Barrier
XJB	Nuclear Waste
RBX	Barrier
CAR 20t Brake	-

7Z96 Winfrith–Gloucester, October 1989	
Locomotive Number: 73119	
Vehicle Type	Load
RNA	Barrier
FNA	Nuclear Waste
RNA	Barrier
CAR 20t Brake	-

7P42 Heysham–Sellafield, June 1989	
Locomotive Number: 31275 & 31130	
Vehicle Type	Load
RNA	Barrier
FNA	Nuclear Waste
RNA	Barrier
CAR 20t Brake	-

7C40 Valley–Sellafield, September 1994	
Locomotive Number: 31201 & 31134	
Vehicle Type	Load
RNA	Barrier
FNA	Nuclear Waste
RNA x2	Barrier
FNA x2	Nuclear Waste
RNA	Barrier
CAR 20t Brake	-

Winscale Flasks seen here at Tyne Yard in 1990. (*David Ford*)

Plumpton Junction, Valley to Sellafield Nuclear Flasks, 1994. (*Dave Peachey*)

Breakdown Trains

Breakdown trains in this era essentially consisted of two main types.

The first type of breakdown train normally contained a tool van which was essentially a travelling workshop with benches, vices, jacks, ropes, ladders and any lifting tackle or winches and one or more 'packing vans' which contained various sizes of timber etc. to assist in the rerailing of errant vehicles. A 'riding vehicle' may also have been included which had cooking and rest areas for the breakdown crew and a compartment/office for the train guard. This type of breakdown train was the most commonly employed and would attend minor through to medium sized derailments which were the most common form of mishap on the network.

Packing and tool vans were normally converted from redundant wagons and coaching stock, up until the mid-1980s most often they would be pre-BR designs. This would include four- and six-wheel vans and coaches, and bogie coaches (usually ex-brakes). In the early 1980s, a number of Mk 1 coaches were specially converted, and the older vehicles were withdrawn. Unlike the cranes which they often operated alongside, packing and tool vans, being derived mostly from ex-revenue stock, had shorter lives, 10 to 15 years being the norm (although the Mk 1 vehicles lasted until the twenty-first century).

The second type of breakdown train is probably the one more familiar to modellers; a steam or diesel-powered crane with a lifting capacity of between 45 and 75 tonnes which would be used to lift large vehicles etc. at the site of major accidents. This type of crane was also sometimes used on civil engineering projects to lift bridge girders and the like into position. The crane would typically be attended by a tool/workshop

Cowans Sheldon 75t Diesel Crane seen travelling in the Toton Breakdown Train 1976. The train is an interesting mix of vehicles converted from pre-nationalisation and ex Mk 1 stock. Note the smoke coming from the heating stove in the guard's compartment at the rear of the train. (*David Ford*)

van, one or more packing vans and a staff/riding coach.

The older types of cranes were powered by steam rather than diesel, so a steam crane would also most likely have a water tanker included in the train. This was often a converted milk tank wagon but other types of tank wagon were also used.

Toton, October 1976	
Locomotive Number: 20176	
Vehicle Type	Load
Cowans Sheldon 75t Diesel Breakdown Crane	-
QQA Tool/Generator Van	Ex Mk 1 Coach
QQV Tool/Packing Van	Ex LNER Composite Coach
QPV Riding Coach	Ex LMS Composite Coach

Breakdown cranes were 'through piped' for vacuum brakes and would normally be marshalled directly behind the locomotive, with the crane jib facing the direction of travel. The packing, tool and riding vehicles would be marshalled behind the crane and would be fitted with vacuum brakes. Upon arrival at the worksite, the line would normally be under an Engineers' Possession (i.e., closed to regular traffic) to allow the crane to work. The tool and packing vans could be detached from the crane and the crane could be moved on its own as required.

Steam powered breakdown cranes were painted red and operated through to the mid-1980s. From the mid-1970s onwards, many cranes were converted to diesel power at Derby locomotive works. Diesel powered breakdown cranes were painted yellow. Attendant breakdown vehicles were painted red, then yellow. It was common to see a mixture of both red and yellow vehicles in the same train until the late 1980s.

Morpeth 1986, 31196 is hauling a single breakdown vehicle (a converted Mark 1 Coach), carrying rerailing equipment. (*David Ford*)

Test Trains

There were a number of train types colloquially referred to as 'test trains'. Such trains could comprise:

- Trains being hauled by newly overhauled or converted locomotives from a BR main works. The purpose of such a run could be to test out locomotive upgrades or modifications or check locomotive performance after heavy overhaul or rebuild.
- Trains hauled on behalf of BR Research. Such trains could be testing new technologies or ways of working, such as experimental suspension systems.
- Trains hauled on behalf of the Mechanical and Electrical Engineers (M&EE). Such trains could be used to inspect the permanent way and infrastructure or the testing and commissioning of new locomotives and rolling stock.

It would be impossible to cover the huge variety of test trains run across BR metals over the years, and thus information should be sought from specialist publications. However, the following list gives a broad indication of the nature and types of train that could be run:

- Ride testing – New or modified suspension or new vehicle types. Often experimental.
- Testing and Commissioning of New, modified, or overhauled Vehicles
- Pantograph Testing
- Ultrasonic or Laser Inspection of Track and Structures
- Vehicle/track interaction and adhesion testing

Prototypes such as the APT-E and APT-P would have been operated as test trains, certainly in their initial guises. Newly manufactured locomotives would also undergo extensive mainline testing prior to being accepted for general BR service. This could result in some very unusual workings.

Seen here passing Wilmorton, 59001 was only two weeks old when photographed undergoing commissioning tests with Test Car 10 in attendance. February 1986. (*Dave Peachey*)

Ex-Class 24 97201 and ex-Class 46 97403 'Ixion' are seen here at the head of the tribology train consisting of Auto Trailer RDB975076, Lab 11 (RDB975046) and van RDB999900. The train made several trips along the (disused) colliery branch testing vehicle/track interaction (rail adhesion). The train is seen departing Tibshelf Sidings for return to the RTC at Derby, March 1986. (*Dave Peachey*)

Not all RTC workings featured RTC stock! Seen here at Stenson, 45110 hauls prototype Class 210 DEMU, 210001, to Derby RTC for further testing. Note the unusual formation, the brakevan at the rear and the translator wagon required to enable coupling between the locomotive and the unit. (*Dave Peachey*)

20131 and 20128, haul the Tribology Train on an RTC-Edwalton working in 1994. The train is seen here passing Wilmorton. Note the revised RTC livery sported by the coaches which was introduced in the late 1980s. (*Dave Peachey*)

Chapter 3

Loads and the Loading of Wagons

The information in this chapter is largely derived from the *British Rail Working Manual for Rail Staff Section 2 for the Loading and Conveyance of Freight Traffic*, the so-called 'Green Pages'. This manual gave instructions to railway staff on how goods and material were to be loaded onto railway wagons.

For modelling purposes, I have only delved into wagons where a load was readily visible, covered vans and the like are not included as the appearance of such vehicles is unchanged on the model whether they are 'loaded' or not.

It must be remembered that 'real life' did not always follow the instructions laid down in the relevant rules, so there are many photographs showing less than perfect adherence to the regulations!

General Rules

There were some general principles applied to the loading and handling of all freight wagons. Some of the rules that are applicable to model railways are listed below.

a) All payloads were required to be adequately secured to prevent them shifting during transit and appropriate securing equipment such as chains, ratchet straps etc. were to be used as applicable.

b) Unless otherwise noted, stanchions on bogie bolster wagons were to be placed as close to the load as possible.

c) When loading goods of unequal lengths, such as timber and steel, the longer pieces were placed at the bottom of the load and used to contain the shorter pieces. Such loads would normally be bound using rope or polyester straps.

d) When mixed loads were loaded the lighter goods were placed in such a way that they would not be damaged by having heavier goods placed on top of them.

e) Where straps came into contact with rough/sharp surfaces they were required to be suitably protected by anti-abrasion sleeves, wood packing or other suitable material.

f) Timbers used to support a load were to be in good condition and nailed to bolsters or the wagon floor. Timbers placed between tiers of goods had to be of sufficient strength and number to support the upper tiers. Steel was never used between tiers.

g) After unloading, all packing used was required to be removed from the wagons or secured, to prevent it falling from a moving train.

h) Wagons with retractable hoods or sliding doors could infringe the loading gauge if the doors or hoods were left open. A wagon was prohibited from being moved in a train with open doors or retracted hoods.

i) Sturgeon wagons were not used for traffic that required the provision of stanchions unless the wagon was fitted with bolsters that had stanchion pockets.

j) Straps were not to be used on traffic where the temperature exceeded one-hundred and fifty degrees centigrade, chains being used instead.

k) When sheeting was used to cover open wagons, 'hollows' in the centre of the sheet were to be avoided as they acted as traps for moisture.

l) Sheets were kept tight and fastened to the wagon, with no gaps. Displaced sheeting could foul overhead line structures or come into contact with other trains.

m) Where it was necessary to use two or more sheets to cover an open wagon, they were overlapped with the overlap pointing towards the rear of the train.

n) Wagons were loaded in such a manner that all wheels were uniformly loaded. Any wagons that could not be loaded in such a way would be classified as an 'exceptional load' and subject to additional operating restrictions.

o) If a load rested on only two bolsters, there would be an overhang of 300mm beyond the edge of the bolster.

p) The carrying capacity and permitted axle weight of a wagon was not to be exceeded.

Pipes/Tubes

Loading methods for this type of payload would depend on the weight, length and diameter of the pipes. If wagons with sides were used (and the load was short enough), the wagons would only require minimal packing to prevent damage to the load whilst the train was moving. When loading larger pipes on flat wagons, specially designed cradles were used, with stanchions used to provide additional support.

There were three permitted loading patterns for tubes and pipes. The first consisted of the load arranged in symmetrical tiers, using cradles or timber packing. Cross members were of sufficient size and number to support the load. There would be a minimum of four to support a

Steel pipes on bogie flat wagon secured with cross ties, stanchions and straps. (*British Rail*)

Steel pipes on bogie flat wagon secured with scotches and straps. (*British Rail*)

Rave and bolster saddles. (*British Rail*)

load. When loaded to this pattern, not more than half the diameter of the top tube/pipe would be allowed to protrude beyond the stanchion.

Pipes could also be loaded with a full base layer, which was scotched to prevent movement and distortion of stanchions. Additional pipes would be loaded in a pyramid pattern and stanchions would be provided on each side of the wagon for support. If the pipes were of different lengths, the shortest ones would be loaded on top. Stanchions would not always be used, depending on the load and the wagon type used.

The third method utilised 'saddles' which could be fitted to the wagon rave or bolster. These were used in the place of stanchions and the load would be strapped directly to the wagon.

When loading pipes in a pyramid, a minimum of two straps would be placed over the second tier from the top. The straps would not be tightly tensioned, but instead allow the top tier of pipes to embed into the second tier down. A minimum of three straps would then be placed over the top tier and then all straps would then be tightened to secure the load.

Ingots and Other Concentrated Loads

An ingot was a bar of metal of a more or less standard size. The most basic semi-finished iron ingot was known as a 'pig'. Iron and steel are dense materials, so care has to be taken that the maximum laden weight of a wagon was not exceeded. Modern pigs of iron, intended for mechanical handling were approximately 15cm wide, 22cm thick and 120cm long and would weigh approximately 300 kg.

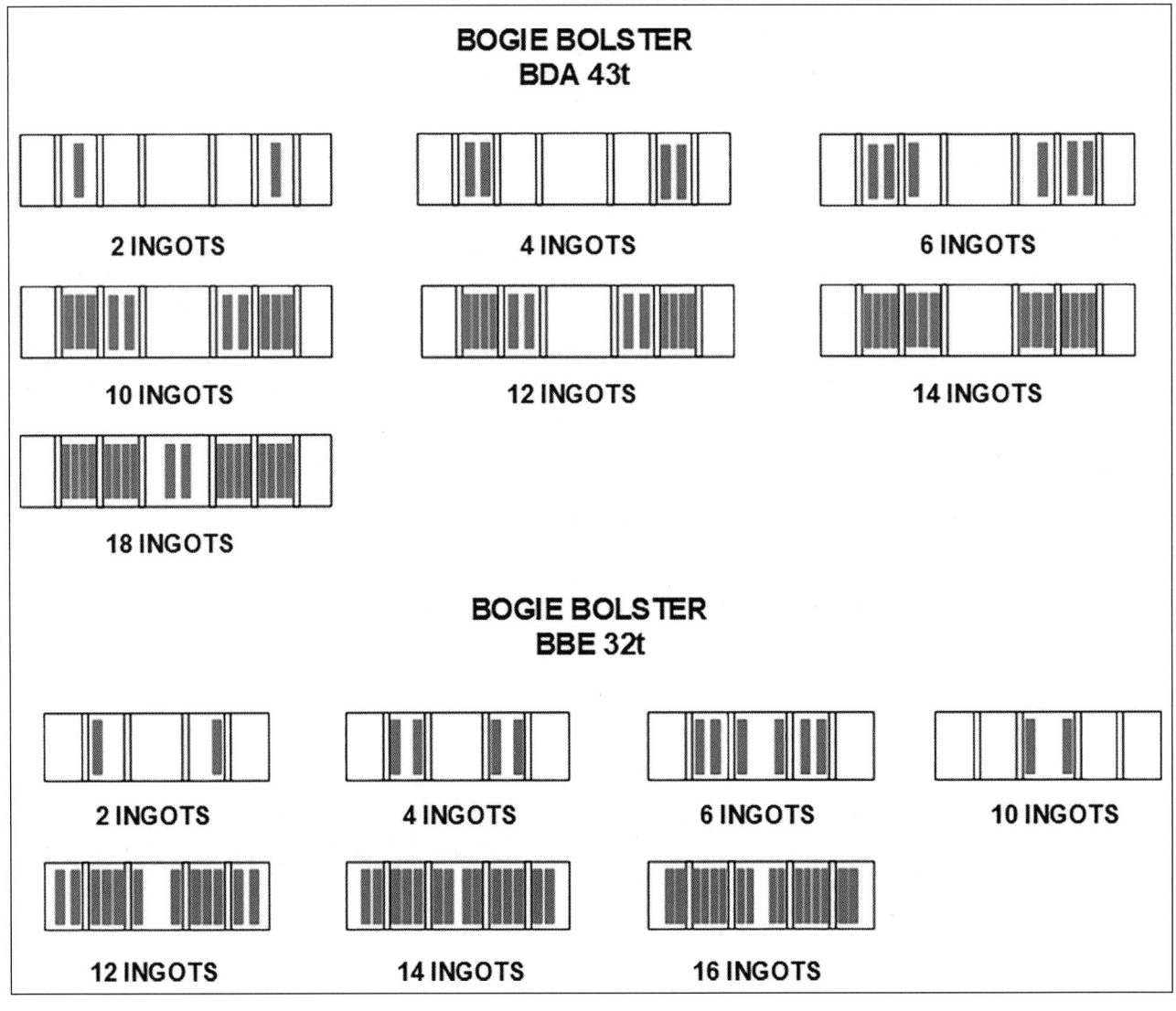

Loading of ingots onto bogie bolster wagons.

A standard 16 ton steel bodied mineral wagon could carry around 50 of the modern pigs, however, in practice they carried much fewer. Purpose built pig iron wagons (TOPS Code UPO), including a batch specifically for 'hot pigs' (TOPS Code URO) were also produced by BR and saw use into the 1980s, their demise finally being caused by the mass withdrawal of unfitted stock in 1985.

By the mid-1980s, carriage of all types of ingots (including pigs) was restricted to air braked bogie bolster wagons (TOPS Code BDA) and air braked steel carrying open wagons (TOPS Code SPA) and followed a very specific loading pattern. All ingots on a single wagon would be of equal size/weight and would be loaded so that the largest area of the ingot would be in contact with the wagon floor. Typically, ingots loaded as part of a block train would not require packing, and simply be stacked on the floor of the wagon.

A billet is effectively a long ingot, typically 15cm square and as long as could be carried by the available rolling stock. Billets were used in wire making and rolled in a steel mill to make small sheets. Due to their length, they would typically be carried on bogie bolster wagons, unfitted and vacuum braked examples being used alongside air braked wagons until the 1980s. A typical load of 15cm square billets on a bogie wagon would perhaps consist of 12-15 billets, weighing approximately one ton each.

Blooms were large raw steel ingots being shipped for further processing. Blooms weighed approximately 30 tons; they were typically around 30cm thick by 76cm wide and around 6m long. Blooms were often loaded 'hot' onto wagons and stock on bloom traffic had raised bolsters inside the wagon to allow cooling air to circulate. Blooms were commonly carried in air braked twin axle wagons, for example the OBA and SPA steel opens which had turn-over or fixed bolsters specifically for this purpose. Blooms were also regularly shipped on air braked bogie steel wagons (TOPS Code BBA) throughout the 1980's.

A slab was a large oblong section length of steel, typically 50cm thick, 60cm wide and

A vacuum braked bogie bolster loaded with billet, seen here at Hereford Yard in 1984. Note the use of stanchions and straps to secure the load. (*Jamerail/Flickr*)

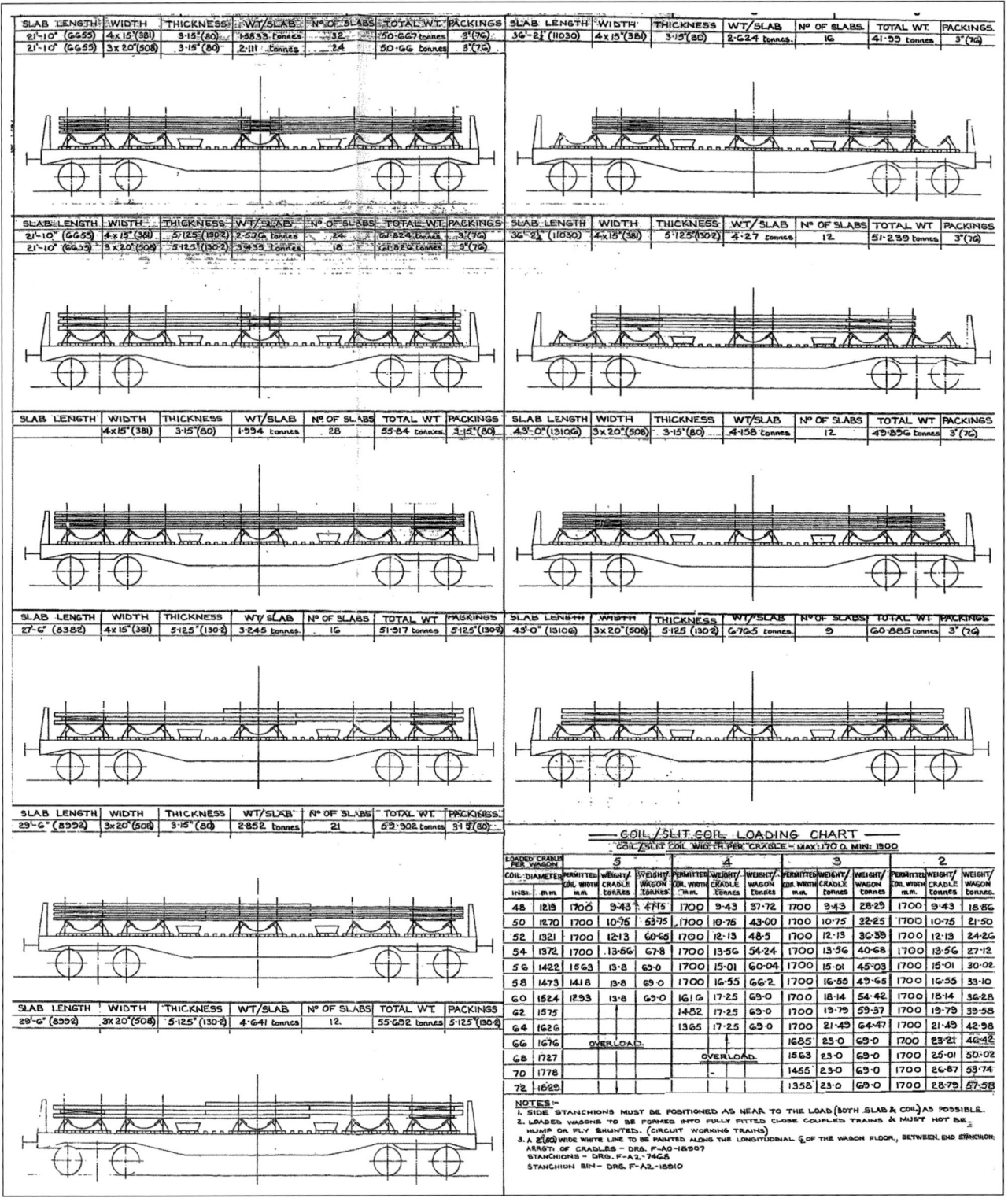

Loading of bogie steel carrier – slab. (British Rail)

around 6m long. Slabs were shipped directly to rolling mills where they were turned into sheet or coil. During the 1980s, many slabs were exported to the continent for rolling, and were then returned to the UK as coil. Wagons carrying such products could often be seen in cross-Channel ferry trains. Slabs were almost always carried on bogie wagons.

Although it would be common for longer and larger ingots to be secured to the wagon either via chains or strapping, it was permitted within the rules to load 'unchained steel' onto bogie bolster wagons, relying instead on the stanchions fitted to the wagon and the weight of the load to provide security during the journey. If steel was loaded in this manner, 300mm of clearance was left between the end of the load and the wagon headstock, the load overhung the outermost bolster by at least 800mm and the height of the load above the bolster was generally not more than 600mm.

Steel – Coil

Strip coil first appeared in the 1950s. By the mid-1970s, strip coils had a diameter of between 1.2 and 2 metres. The larger the diameter, the shorter the roll, as coils were designed to all be the same approximate weight. All the coils in a block train would normally be the same diameter, however a mixed wagonload freight or Speedlink train might consist of wagons carrying different consignments, so mixed sizes were common, particularly from the 1980s onwards.

The coils were almost always banded with metal strips, two or four straps passing through the centre of the coil, and two or three wrapped around the coil to prevent it coming unrolled.

From the 1970s onwards, the smaller older types of coil carrier were being replaced with bogie wagons, which could carry a maximum load of five coils. Most of the unbraked and vacuum braked coil carriers had been withdrawn by the mid-1980s.

If a wagon was to be partially loaded, then the coils would have to be configured in a specific manner to ensure that the wagon was evenly loaded.

Example of wrapped steel coil. (British Rail)

Some coils were imported from Europe and often carried on twin-axle air braked open wagons such as the OCA and SPA. The coils would be fitted to forkliftable cradles and often wrapped in clear sheet to protect them from corrosion. Two coils per wagon was a standard load.

Coil was generally loaded onto a wagon in one of three configurations:

a) Axis transverse – i.e., 'eye to side'
b) Axis vertical – i.e., 'eye to sky'
c) Axis longitudinal – i.e., 'eye to end'

When loading coil, it was also important to make sure that the load was spaced equally across the wagon bolsters, otherwise the wagon could be dangerously unstable which could result in a derailment.

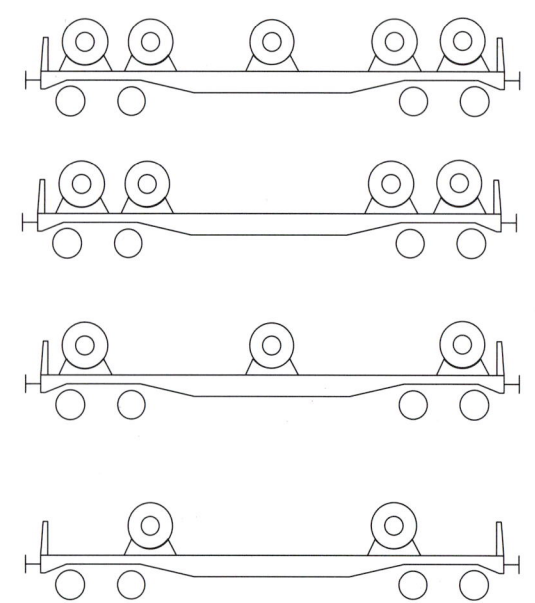

Correct loading of coil onto bogie steel carrier showing even distribution of load.

Strip Coil on Air Braked Bogie Wagons loaded (from Left to Right, Top to Bottom) 'Eye to Sky', 'Eye to Side' and 'Eye to End'. (*Jamerail/Flickr*)

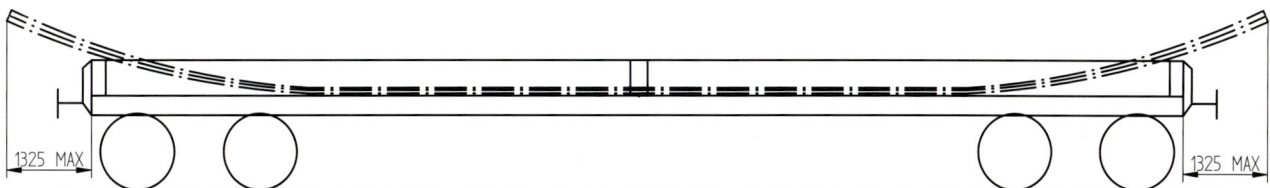

Loading of flexible plate on Boplate type wagons.

Steel – Sheet and Plate

Steel plate and steel sheet were commonly carried items on the railway. These larger loads were normally carried on vacuum or air braked bogie bolster wagons, but wagons without bolsters such as plate wagons were also used, the load being supported on wooden crossbars if required. Plate or sheet that could fully fit inside an open wagon would be spaced equally over the floor, and the top of the load was not permitted to be higher than 100mm from the top of the wagon rave.

Thin gauge flexible plate was sometimes simply placed in the bottom of an open wagon such as a Boplate. An amount of overhang was allowed at each end of the wagon as long as it was equidistant across the wagon, this was typically a maximum of 1325mm from the wagon ends.

Sheet or plate was not permitted to be loaded onto the floor of wagons with end doors.

Sheet and plate on bolster wagons would typically be secured by chains or ropes, passing under the load and being secured to the wagon tie down points. Thinner gauge sheet might simply be placed in the wagon, using the wagon ends and sides to contain the load.

Heavy steel plate, typically over 8ft in width would have to be carried at an angle on 'Trestrol' wagons which incorporated a specially adapted trestle frame.

Large metal plates were uncommon by the 1970s, mainly due to a significant decline in heavy industry such as ship building. However, wagons carrying large plate could still be seen around iron/steelworks, shipyards, and other heavy engineering facilities.

Steel – Rails

Rails would normally be carried for engineering and maintenance purposes, but they might also be carried as general traffic, for transport to specialist assembly locations for machining and incorporation into switches and crossings, for example. The Thomas Ward works at Sandiacre regularly received such traffic until the late 1990s. It would also be common to

As seen on 'Shirebrook', this photograph shows N Gauge Salmon wagons loaded with lengths of rail. (*Duncan Hunnisett*)

Diagram E3
Loading Bullhead Rails - Nested.

(vi) 42.5t wagon: 50 rails in five tiers.

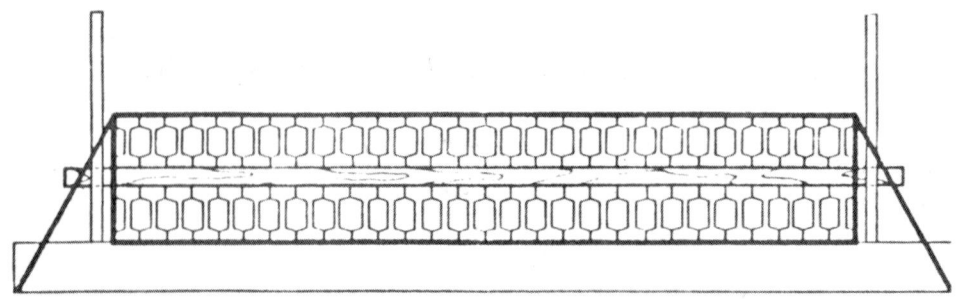

Loading pattern for nested bullhead rails. (*British Rail*)

Diagram E3
Loading Bullhead rails vertically on YLA and BDA/BDW wagons

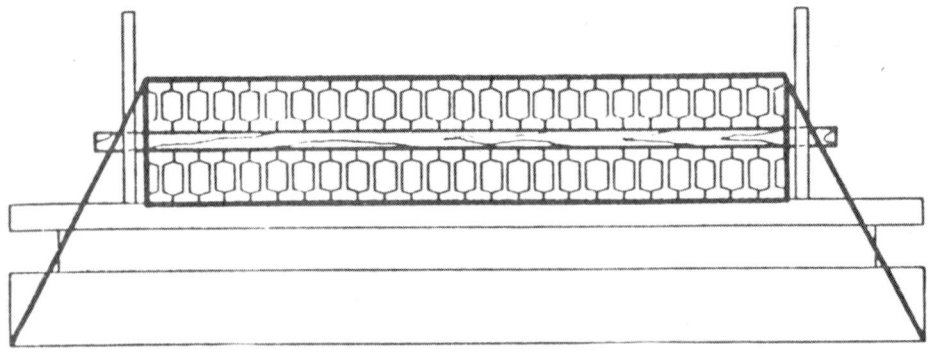

(iv) 49t wagon (YLA): 60 ft rail length.
56 rails in two tiers, heads up.
75 mm x 50 mm timber packings between tiers extending 225 mm beyond rails, and to be positioned at least 600 mm from side stanchions.
Rails to be strangle wrapped at ends.
Stanchions to be set at 2006 mm centres.

(v) 42.5t wagon (BDW): 60 ft rail length.
48 rails in two tiers, heads up.
75 mm x 50 mm timber packings between tiers extending 225 mm beyond rails, and to be positioned at least 600 mm from side stanchion.
Rails to be strangle wrapped at ends.
Stanchions to be set at 1752 mm centres.

Loading of bullhead rails vertically on YLA and BDA/BDW wagons. (*British Rail*)

see newly manufactured rails in block trains being transported from the mill, the British Steel works at Workington, for example, was a prodigious producer of rail.

Lengths of rail under 18.5m in length were commonly carried on bogie bolster wagons, supported by stanchions and strapped down. The height of rails loaded in this manner did not exceed 4 rails. Thin timber baulks were inserted between each layer of rails to prevent damage to the rail head.

Bullhead type rail could be loaded 'nested' owing to its unique geometry. This did not require timber baulks or packing and could be directly strapped to the bed of a wagon up to five tiers deep. On a typical air braked bogie wagon, fifty such rails could be carried.

It was also common to load bullhead rail in tiers as per standard rail but owing to the fact it only has a thin foot, and the danger of toppling, the rail was normally only loaded two tiers deep. Between 48 and 56 rails could be carried on a typical flat wagon in this manner.

Rails over 18.5m in length, which became increasingly common as lines moved over to CWR, required specific loading to prevent

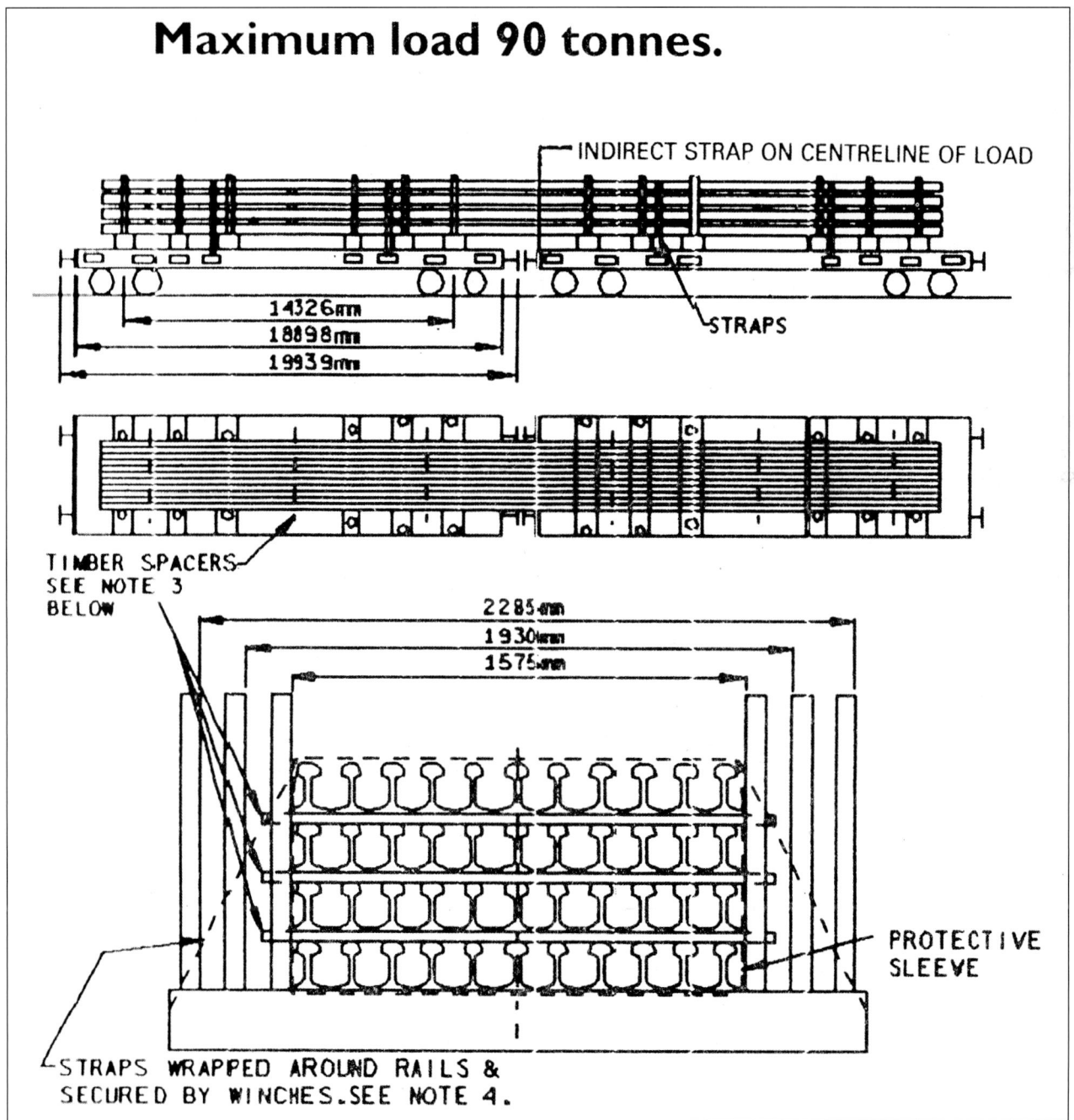

Loading of rail in excess of 18.5m to YLA mullet wagons. (*British Rail*)

An unidentified Class 37 hauls a rake of bogie flats loaded with lengths of CWR rail towards Redhill Tunnel in 1987. (*Jamerail/Flickr*)

damage. This loading would take into account the natural flex of the rail, and thus the load could be carried across multiple wagons. Specific pools of engineers' wagons such as the YLA 'Mullet' and YAA 'Brill' air braked bogie wagons would often be used in multiple to carry long lengths of rail. There was also a special train comprised of 'Perch' wagons (TOPS Code YEA) with a special mechanical handling unit specifically for the loading, transportation and unloading of large amounts of welded rail.

Scrap Metal

Scrap metal was most commonly conveyed in open wagons, either wagons specifically designed for the traffic, or repurposed wagons such as ex-traffic opens or hopper wagons.

Large pieces of scrap would be positioned in the lower part of the wagon, i.e., not above the line of the rave. Loose scrap would not be loaded above the rave of the wagon.

If the scrap was in bales or bundles that exceeded the height of the wagon that they were being loaded onto, they would be arranged in such as manner as to create a stable and secure load which would not shift in transit.

Scrap was not permitted to be loaded above the rave of a wagon in wagons with a rave height in excess of 3350mm.

Timber

In the early part of the twentieth century, most timber carried on the railways was tree trunks and long heavy baulks, the cutting, planing and finishing of the wood was done at the local timber merchants or in the factory which used the wood.

Timber is generally felled in late autumn when sap is at its lowest level in the tree, the leaves have fallen, and the general ground cover is reduced making it easier to drag the trunks out of the woods. Timber can also be cut in the midsummer, again when the sap is at a low ebb, although recovering the wood is then more difficult. Tree trunks, tree stumps and timber large baulks went on flat wagons, well roped down. Shorter pieces of timber (such as pit props) were often carried in open wagons.

Loading of Scrap in Open Wagon. (*British Rail*)

OTA wagon loaded with timber. (*British Rail*)

A great deal of domestic timber traffic came from the cultivated forests of Scotland, and by the late 1980s, this traffic was mostly handed by air braked timber carriers (TOPS Code OTA).

The UK also imported a great deal of pre-cut timber, usually Baltic softwoods (such as pine) from Scandinavia and Canada. This was colloquially referred to as 'deal' timber, (deal was an old word for cut pine).

Obviously with this being an imported product most traffic flows were from the docks, but the destinations could be anywhere on the system. Close to docks and ports entire trains of such timber heading for the marshalling yards could be seen.

Open wagons were used, but also, flat wagons were used for longer lengths. A number of ex-Cargowaggon vans were converted into timber carrying wagons for long lengths of imported timber.

Mines used large quantities of wooden pit props; during the 1920s, about eight million tons of pit props were moved by rail a year but this fell precipitously as the UK mining industry wound down. However, well into the 1980s, there were still large numbers of collieries in the UK, and thus they would need regular deliveries of pit props.

Pit props were always softwood, and were supplied in lengths ranging from 0.5-3.5m and in diameters ranging between 7 and 60cm.

Pit props were always used with the bark removed and by the 1980s nearly all props

ex Cargowaggon chassis loaded with timber. (*British Rail*)

OBA and OCA wagons loaded with timber, Hereford 1985. (*Jamerail/Flickr*)

were shipped in this condition. The props were generally straight and roughly even in length, with no protruding knots or branch stubs which might be snagged and cause a collapse.

'Split props' were props which were cut along their length and had a half round cross-section. Split props were used to support the roof, being supported in turn by conventional props at either end. Wooden pit props were still used in British coal mines in this era, although only for temporary support. It would be usual to see several wagon loads in a train, the props were generally imported, and would most often be delivered as a block load.

Another essentially similar load regularly hauled from the docks would be 'pulp wood' destined for large paper works. This load was similar in size and appearance to pit props and shipped with all bark removed. Such loads sometimes featured in Speedlink trains.

If the props were short, 2m or less, stacks of props could be stacked up in a suitable open wagon. This generally meant the props were sticking out over one end of the wagon and a runner wagon would be placed at the end of a rake of such wagons so the overhanging load of the last wagon would not foul any vans which might be adjacent in the train.

Longer props could be stacked in this manner if a suitable long wheelbase open wagon, such as an OBA or OCA, were used.

Bogies and Wheelsets

Although rarely photographed, the transfer of bogies and wheelsets between main works and depots and yards was fairly common. BR created the ENPARTS service to move spares and components between works and depots, and a variety of wagons and coaches, both dedicated and ex-traffic were used.

Wheelsets could be transported on bogie bolster or in open wagons. Bogies (with and without wheelsets) would normally be transported on bogie flat wagons, although occasionally a 'well' or trolley-type flat wagon might be utilised. Bogies and wheelsets would be loaded onto timber supports which extended the length of the wagon floor.

The wheelsets would be loaded in a specific manner to avoid damage, and in all circumstances, steps would be taken to ensure that the flanges of adjacent wheels could not impact one another (which could cause damage). Each wheelset would be individually secured to the wagon to prevent movement.

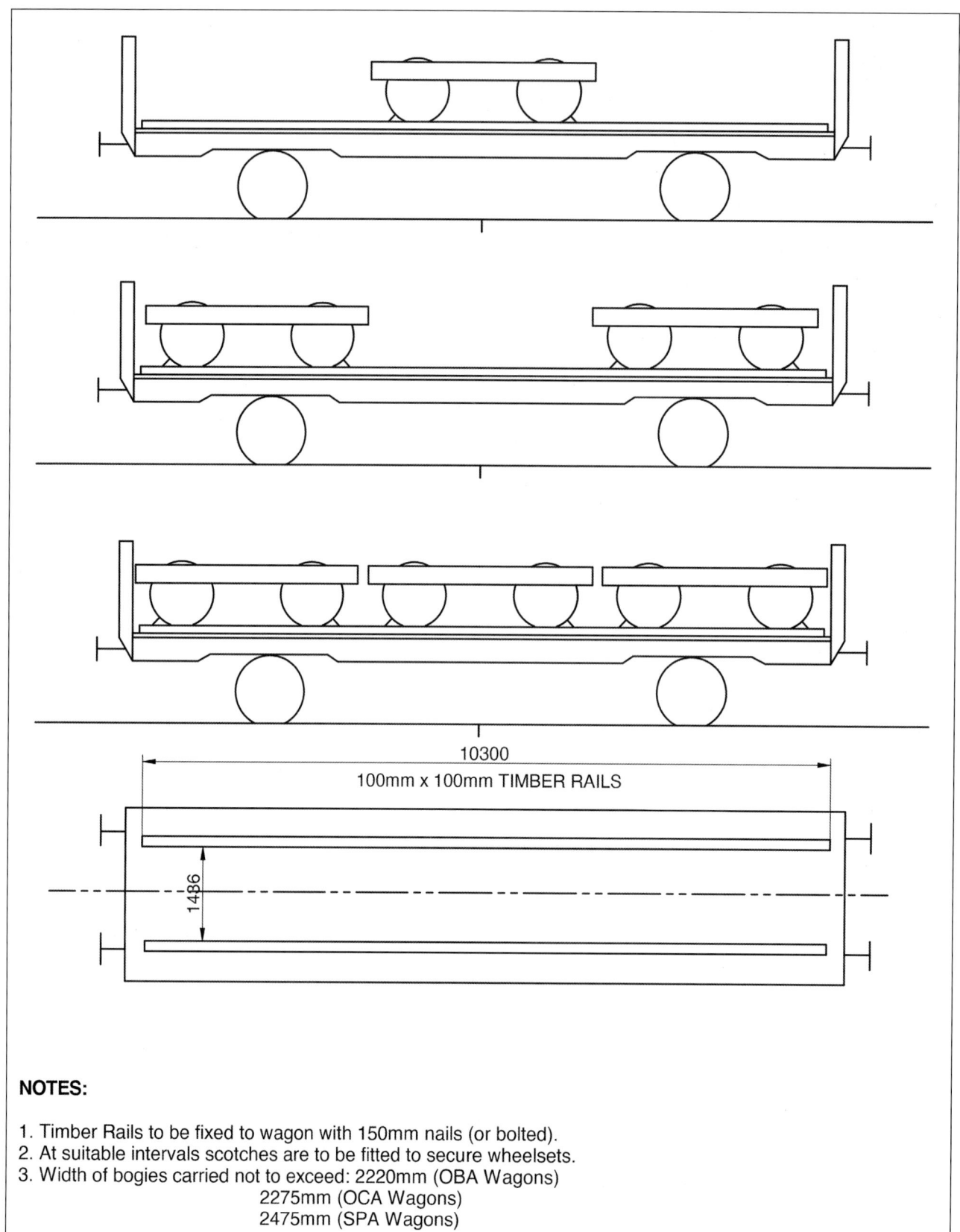

NOTES:

1. Timber Rails to be fixed to wagon with 150mm nails (or bolted).
2. At suitable intervals scotches are to be fitted to secure wheelsets.
3. Width of bogies carried not to exceed: 2220mm (OBA Wagons)
   2275mm (OCA Wagons)
   2475mm (SPA Wagons)

Loading of Bogies.

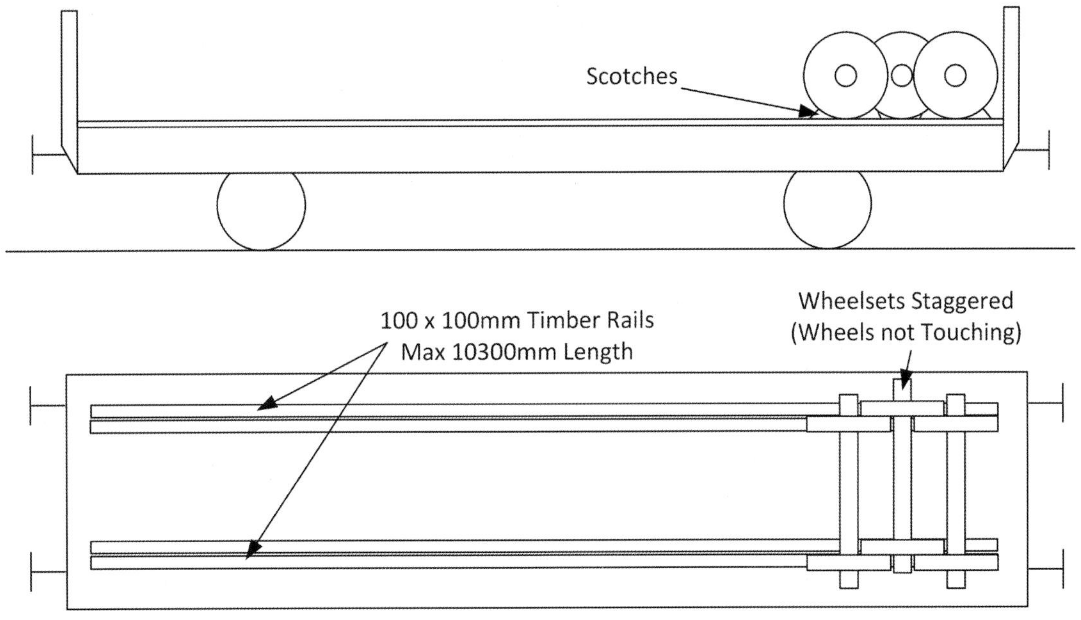

Scotches

Wheelsets Staggered (Wheels not Touching)

100 x 100mm Timber Rails Max 10300mm Length

NOTES:
TIMBER RAILS TO BE FIXED TO WAGON FLOOR
ALL WHEELSETS TO BE SCOTCHED IN FRONT AND BEHIND EACH WHEEL
WHEELSETS TO BE UNIFORMLY DISTRIBUTED ALONG WAGON LENGTH
OVERALL WIDTH OF WHEELSETS TO BE CARRIED TO NOT EXCEED:
2075mm for OBA Wagons
2130mm for OCA Wagons
2330mm for SPA Wagons

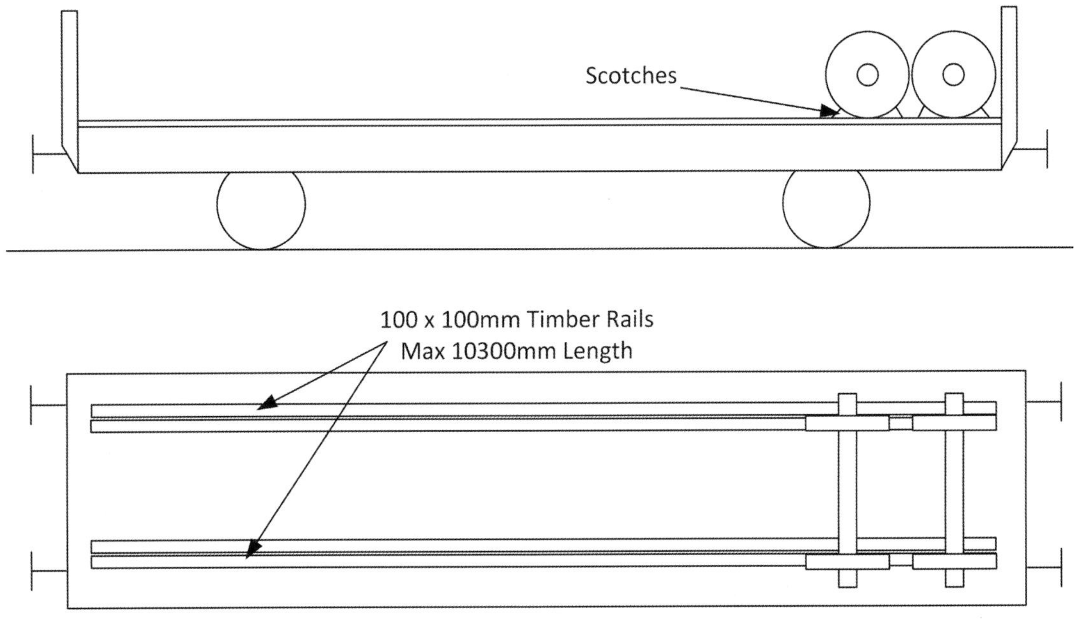

Scotches

100 x 100mm Timber Rails Max 10300mm Length

NOTES:
TIMBER RAILS TO BE FIXED TO WAGON FLOOR
ALL WHEELSETS TO BE SCOTCHED IN FRONT AND BEHIND EACH WHEEL
WHEELSETS TO BE UNIFORMLY DISTRIBUTEED ALONG WAGON LENGTH
OVERALL WIDTH OF WHEELSETS TO BE CARRIED TO NOT EXCEED:
2200mm for OBA Wagons
2275mm for OCA Wagons
2475mm for SPA Wagons

Loading of Wheelsets.

Reeled Paper

Although not a particularly common traffic, reeled paper continued to be moved by rail for many years, even into the 1990s. Paper reels are large, and would either transported in open wagons, or, in later years, covered vans.

When used to move reeled paper, open wagons would generally be clean and in good repair, with dry floors free from protruding items such as nails which could damage the product. The floor was often covered in old wagon sheets or tarpaulins to keep the outer coverings of the reels clean and dry.

Reels were loaded directly into the wagon and prevented from movement by rubber pads or similar packing materials, which were placed between the wagon ends and the reels. If the reels were to be loaded more than one tier high, the subsequent tiers were firmly nested in the saddles of the lower tiers to prevent movement. Up to three tiers could be loaded in certain circumstances with the top tier reels partially above the rave of the wagon. If the reels did not completely fill the bottom of the wagon, a 'key' reel was used wedged between the end of the wagon and the reels to secure the load.

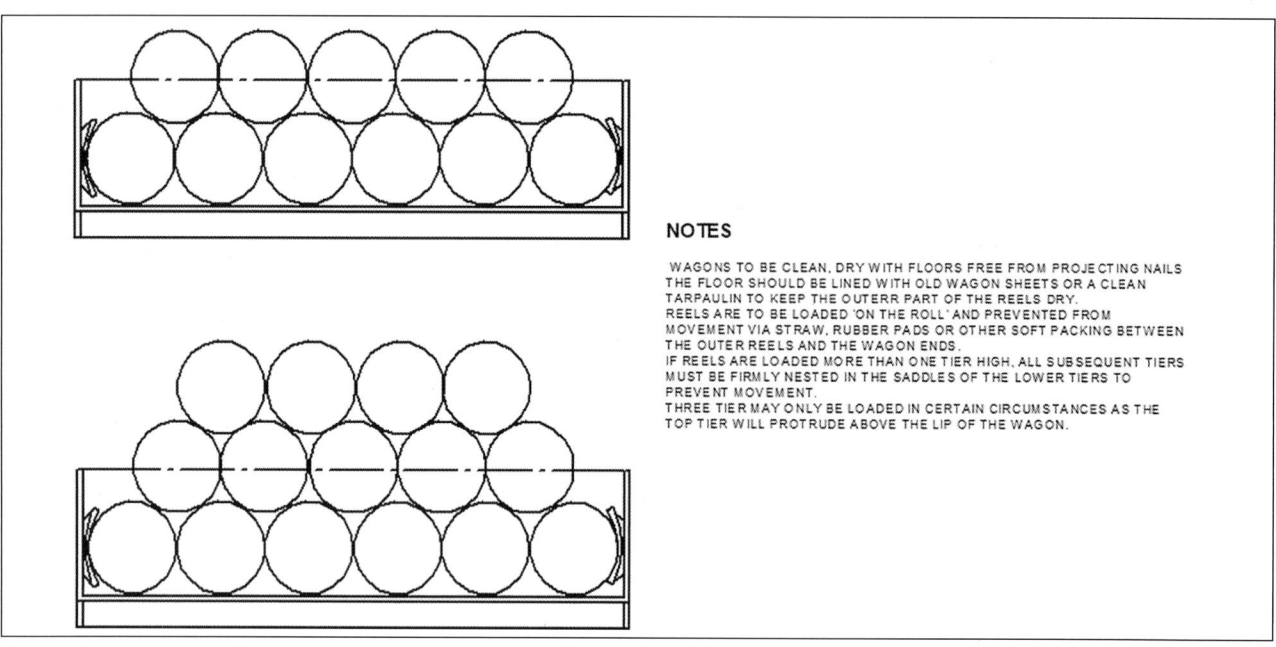

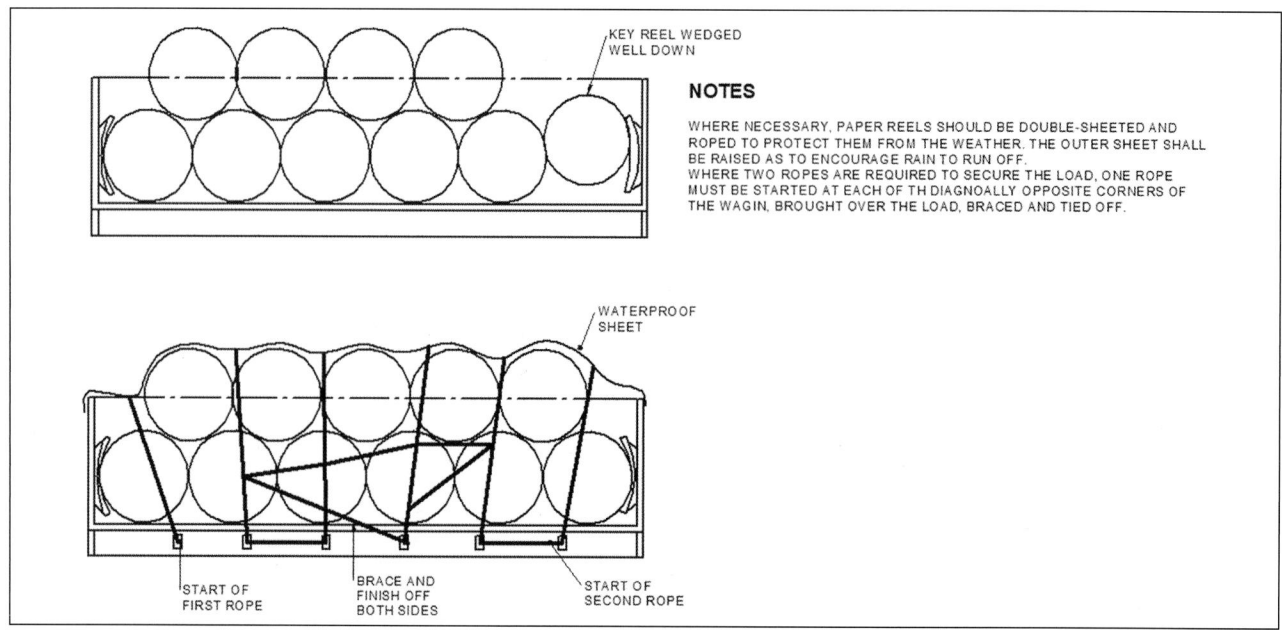

Loading of reeled paper in open wagons.

Loads of paper reels were generally doubled sheeted for rain protection and roped. The outer sheet was raised so that the water would run off.

Road Vehicles and Plant

Large vehicles such as tractors, excavators, bulldozers and so on would foul the loading gauge if loaded onto standard flat wagons, so a special type of wagon known as a 'well' or 'machinery' wagon was developed. The design has pre-BR origins, and BR inherited a number of pre-nationalisation machinery wagons, and even built some of their own, often to pre-nationalisation designs. The basic idea of such a wagon is that the centre of the wagon (the well) is lowered. Therefore, when the wagon is loaded with a large load the load remains 'in gauge'.

There were a significant number of well wagons of different types, but by the 1980s most of the remaining examples of these wagons had been transferred to the engineers, where they were useful due to their ability to carry heavy plant.

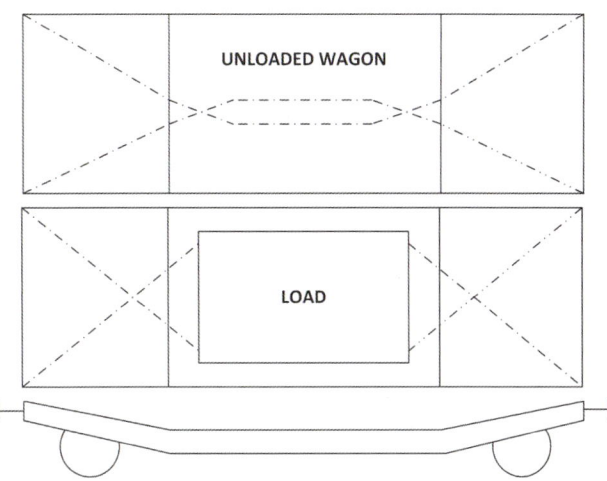

Machinery/well wagon loading details.

The load would be secured on each corner via a chain (or chains might be laid over smaller loads with suitable protection). When the wagons were unloaded, the chains would often be left attached, but 'stowed' in a tidy manner. Vehicles would often also be chocked to prevent movement.

Some 'Warwell' wagons were also used in general traffic.

An unidentified Flatrol loaded with BR tracked excavator, seen here at Wooferton in 1987. The machine is strapped to the wagon via straps passing under the excavator bucket. (*Jamerail/Flickr*)

A Typical ex-LNER 'Lowmac' (TOPS code ZVW) seen here at Cardiff Tidal Sidings in 1988. Note the tidy manner in which the chains have been left. (*Jamerail/Flickr*)

Motor lorries and other outsize vehicles would sit too high and foul the loading gauge when loaded onto standard height wagons. Smaller vehicles might be loaded onto a bogie well wagon, with the vehicle's front wheels in the well and the rear wheels on one raised end which would help to bring the vehicle inside the loading gauge.

In the mid 1980s a single decked version of the 'Autic Six' car wagon appeared. These were known as 'Comtics' and they were used for notably of exports of cab and chassis units from Leyland in Lancashire. The wagons would normally carry three lorry cabs, one in each dropped bay and a third loaded over the joint in the centre.

Typical loading of 'Comtic' type wagon. (*British Rail*)

Pre-Assembled Track Panels

Track panels for renewal works would often be made up at Pre-Assembly Depots (PADs) and then loaded onto Engineers' wagons for transportation to the work site. Panels were normally limited to four deep.

Old track panels, removed during renewal work and perhaps required for reuse in sidings and yards, would be transported onto to their destination if required again using available engineers' flat wagons.

Right: This photograph taken at Cramlington in 1984, illustrates how track panels were typically loaded. (*David Ford*)

Below: Securing of pre-assembled track panel on flat wagon. (*British Rail*)

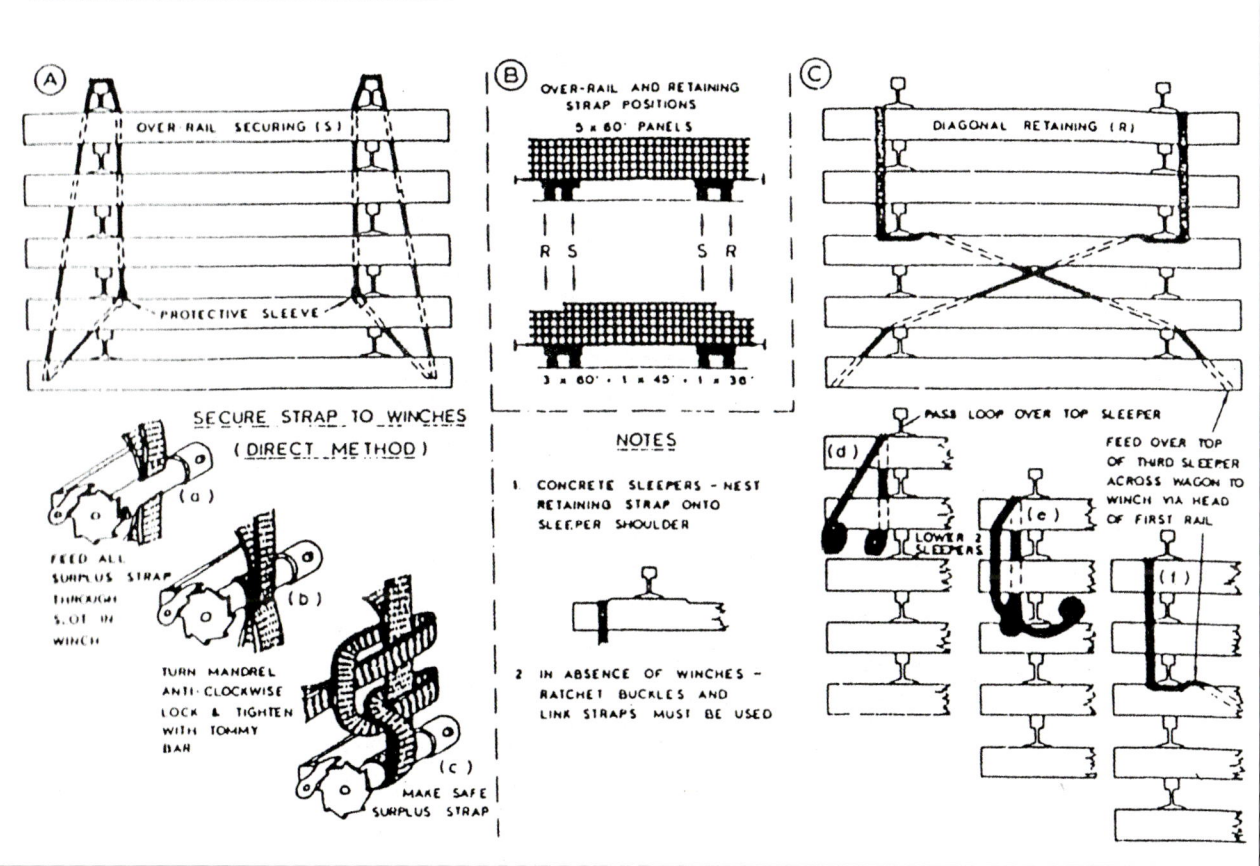

Sleepers

New sleepers would be required for major works, and these were often delivered to the worksite using specially modified wagons; during the 1970s, vacuum braked Borail bogie flats were often used. These were upgraded with air brakes in the 1980s and were known as Mullet or Parr wagons. Smaller quantities of new sleepers might be transported loose in an open wagon, various forms of which were used by the engineers, including redundant air braked wagons such as the OCA and SPA. Old sleepers would also need to be transported away from site, and this was normally carried out in a similar fashion, the difference being that old sleepers would typically be thrown into the wagon haphazardly rather than carefully loaded onto specialist wagons

The 'stack' on a wagon such as the 'Parr' would be limited to a maximum depth of three sleepers deep, carrying a maximum of 168 sleepers per wagon. The stack would be secured to the wagon using ratchet straps, and longitudinal beams would be fitted on the top

Closeup of ratchet straps used to secure sleepers on Parr sleeper wagon. (*British Rail*)

A Parr wagon loaded with concrete sleepers stabled at Gresty Lane, Crewe in 1993. (*Jamerail/Flickr*)

OBA and SPA wagons in engineer services loaded with wooden sleepers in Hereford Yard. Also note the ex-milk tankers loaded with water and a Borail wagon carrying concrete sections, 1985. (Jamerail/Flickr)

of the load to prevent individual sleepers from moving during transport. Smaller numbers of sleepers, commonly timber sleepers but also sometimes concrete and steel examples, required for minor renewals work, would be tidily stacked inside an open wagon and would not normally be secured with strapping as the sleepers would be retained by the wagon sides.

Cable Drums

Loaded cable drums were carried in high-sided goods wagons wherever possible. They were always loaded 'on the roll', i.e., not flat, to prevent tangling of the cable. Drums up to around 900mm in diameter were loaded with the axis of the drum either longitudinal or transverse, according to the number of drums to be loaded into the wagon. Where required, the drums were secured with wooden or rubber scotches or packing.

Drums larger than 900mm in diameter were loaded without the use of securing ropes. The typical methods of loading involved securing wooden packing and wooden cradles to the floor of the wagon using nails. If larger wagons were used, multiple larger drums could be loaded in a similar manner.

It was permitted to load drums of varying sizes in the same wagon. Cable drums with a diameter exceeding 2m were classified as a special load and dealt with accordingly.

Ex LNER tube wagons in departmental service loaded with cable rolls at Westbury in 1983. As can be seen, it appears loading instructions have not been rigidly followed, but as most of the drums are empty, and some are marked 'damaged' it is likely these are scrap/surplus from a completed job. (Jamerail/Flickr)

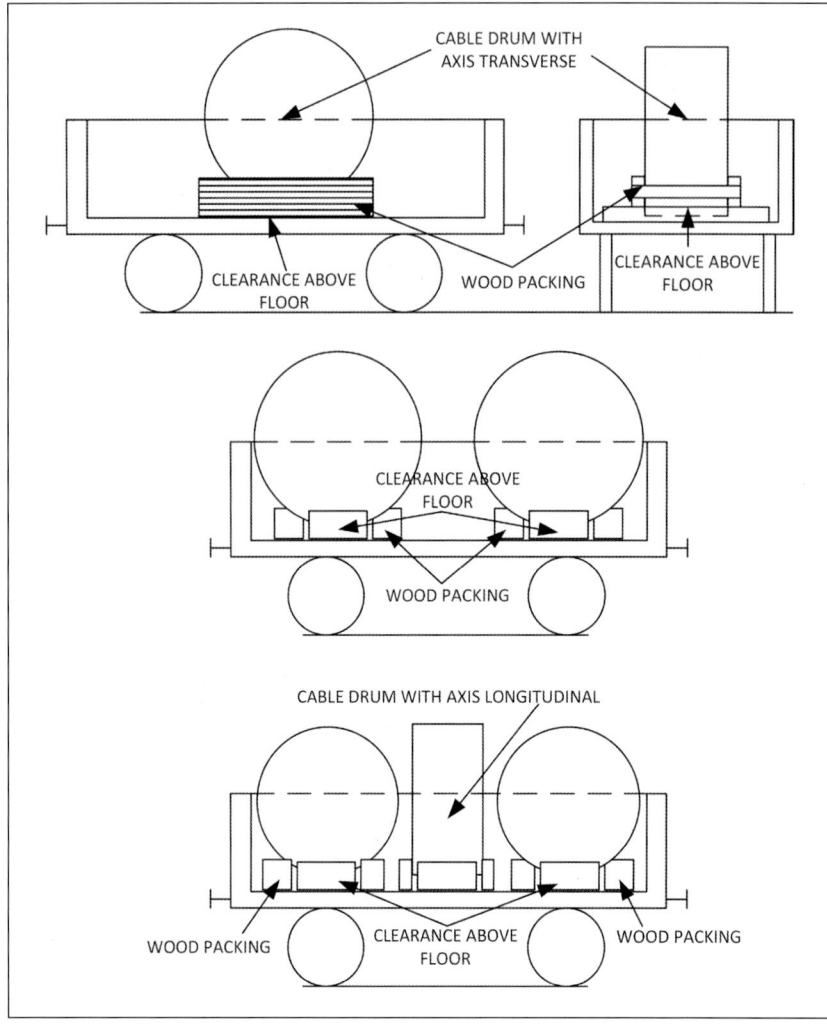

Left: Loading of cable drums in open wagons.

Below: This photograph taken at Newport in 1982, shows an Engineers' wagon loaded with part used cable drums. Note the loading of the drums, they have been arranged so as to form a 'wedge' ensuring they cannot move within the wagon during transit. (*Jamerail/Flickr*)

Spoil and Ballast

Spoil (excavated waste material) and ballast (old and new) were frequently transported. Major works would justify a dedicated spoil train, smaller jobs might require one or two wagons dedicated to ballast or spoil in a general engineering train.

For spoil, almost any open wagon would be used, often an ex-traffic mineral or open wagon, with ex-16 ton mineral wagons finding particular favour from the 1980s onwards. To prevent overloading, mineral wagons often had rectangular holes cut in the sides, as spoil and ballast is a denser material than coal, meaning a heavier weight can be carried in the same space.

New ballast was normally carried in specialist wagons, hopper wagons with centre or side chutes, or side-tipping wagons called

Ex-16 ton minerals in use as spoil wagons seen here in a short engineers' train at Pontrilas in 1987. Note the white rectangles which highlight the bodyside 'cut outs' to prevent overloading of the wagon. (*Jamerail/Flickr*)

Ex-LMS Vacuum-braked 3 Plank open (TOPS Code ZAO), now unfitted, is seen here in Engineers' service loaded with spoil at Doncaster station in 1981. (*Jamerail/Flickr*)

Above: Another photograph taken at Doncaster in 1981 shows an ex LMS 3 plank vacuum braked open wagon in Engineers service, this time loaded with building sand. (*Jamerail/Flickr*)

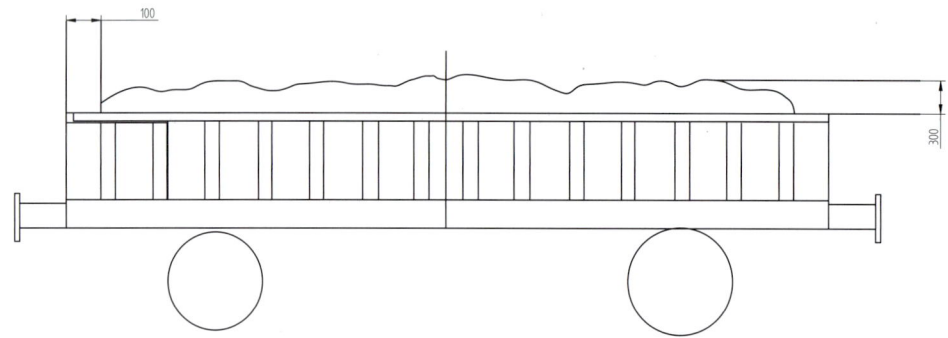

Left: Ballast loading maximum dimensions (MHA Wagon Shown).

'Mermaids'. A train with some wagons might also include a specialist 'Shark' ballast plough. These ploughs, superficially resembling a brake van, distributed the deposited ballast evenly about the rails and reduced the amount of manual work required by the Permanent Way gang.

Old ballast would often be carried in purpose-built engineers wagons such as the 'Grampus' or 'Starfish' types, although it wasn't uncommon to find an ex-traffic open wagon being used. Such wagons often had any automatic brakes removed to simplify maintenance, as engineers trains ran at low speeds and thus it made little difference if the wagons were unfitted.

For wagons carrying old or new ballast, the 'peak' above the wagon 'raves' (the top of the wagon side and ends) was allowed to be approximately 300mm. A recess of approximately 100mm was given around the sides of the wagon; the purpose of this recess was to allow any loose ballast from the peak to roll down but remain in the wagon. The primary purpose of these dimensions was to ensure that the wagon was not overloaded and also to ensure that the ballast (largely) remained in the wagon and was not loaded in an unsafe or unbalanced manner. A review of photographs of loaded ballast wagons suggests that such instructions were not always strictly adhered to!

Loose Materials

Loose materials can cover a variety of different items, scrap, pipe work, trunking parts, signalling and infrastructure parts, old sleepers etc. Generally, there were no specific loading

Ex 13t steel Medfit, coded ZAO with a load of drainage pipework and parts seen here at Doncaster in 1981. (Jamerail/Flickr)

rules for such items. Loose materials loaded into open wagons shouldn't have extended above the wagon sides, but as can quite readily be ascertained from photographs, this was sometimes ignored.

Coal

By the 1970s, coal hoppers were most often loaded by rapid discharge bunkers, which gave the load a distinctive 'pyramid' top. Many locations also had a crude device for ensuring the wagons weren't overloaded, basically a length of plastic or rubber suspended over the track under which the wagons were drawn.

Coal is not a particularly dense material, compared to aggregates, for example, which is why ex-16-ton mineral wagons used for hauling old ballast and spoil had 'letter box' holes cut halfway up the side, to prevent the wagon being overloaded. The coal itself came in various grades, from lumps about the size of a pebble to large 'slabs' which could exceed a foot in length.

Loaded HAA MGR hopper seen here at Doncaster in 1981. (Jamerail/Flickr)

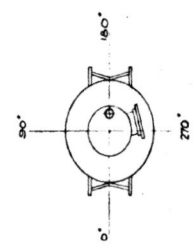

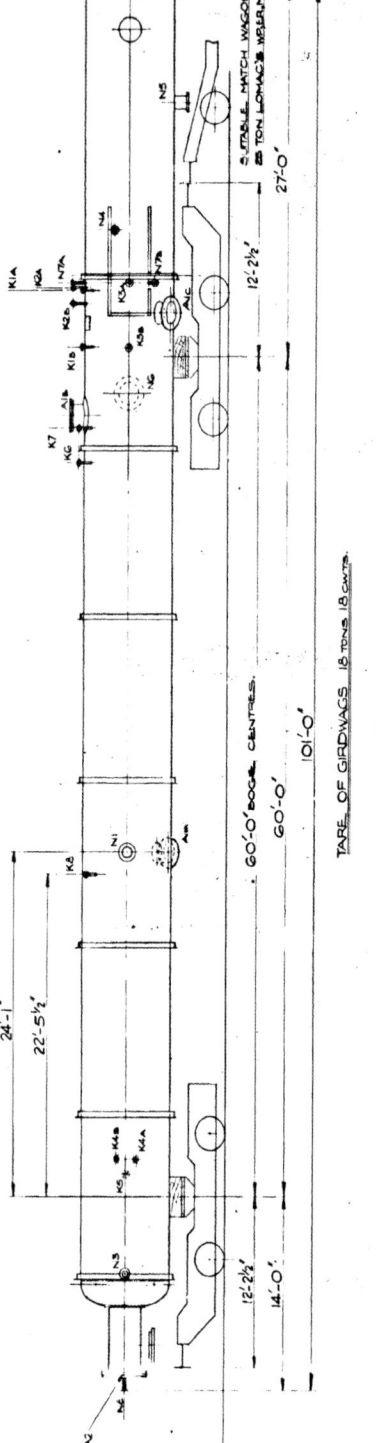

Loading diagram for splitter column loaded onto XXO 20.5 ton girder wagons. (*British Rail*)

Coal would generally be graded either at the pit, or at a rail or road-served screening plant. It would be possible for a rake of wagons to have different grades of coal in them, apart from MGR workings which used a uniform grade of fine coal for use purely in power stations.

Special Loads

The movement of large out-of-gauge loads, such as large girders, cast bridge pieces, splitter columns and so on, required very careful planning and execution. Moves were made at very slow speed and often overnight. Principally, the load would be classified as 'self-contained' if the load would fit entirely on the wagon without any overhang.

The load might also be classified as requiring one or more runner wagons at either end of the wagon to which the load was secured if there was an overhang at one or both ends of the wagon. The purpose of these wagons was to 'contain' the overhang of the load and prevent it striking any adjacent vehicles in the train. The load would not be secured to the runner wagons. If the load was too long for the wagon, a runner would be provided at both ends and there would be equal overhang over both wagon headstocks.

For extremely large or bulky loads, whose diameter was considered outside of the normal loading gauge, a special diagram was prepared by the engineering department giving very precise dimensions, not only for the load, but also the manner in which it was to be loaded to the wagon(s) and the exact type and disposition of any attendant runners. Such dimensions and loading particulars were vital to ensure that the load did not foul any structures en route. These 'out of gauge' loads always travelled with a BR load inspector in an attendant brake van or perhaps riding in the rear cab of the locomotive in later years.

The example left is an extract from a drawing for the transfer of a large splitter column from its factory to Ellesmere Port for export. The wagon in question is a pair of 20.5 tonne girder wagons, coded XXO. A 'Lowmac' machinery wagon was used at the overhang end of the load as a runner wagon.

LOADS AND THE LOADING OF WAGONS • 105

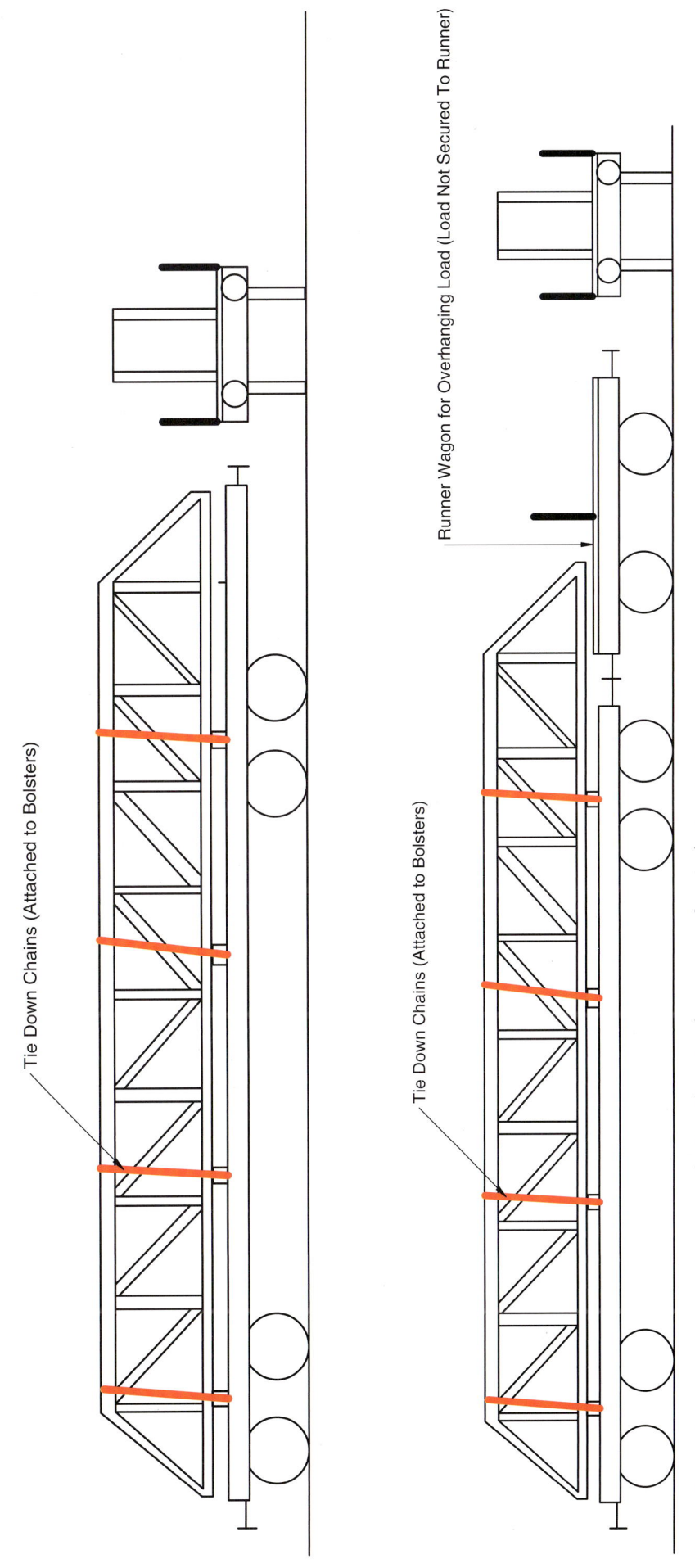

Self Contained Loads without and with overhangs, showing disposition of typical runner wagons.

Chapter 4
Civil Engineering and the Permanent Way

The Permanent Way

Whilst the majority of modellers would probably regard the train as the most important feature of a layout, there is no escaping the fact that the track and associated equipment will form a significant part of the overall scene.

This section will discuss the 'Permanent Way', that has, as its constituent parts, a combination of rails, sleepers, ballasts, fixtures, and fastenings. This is used in railway terms to distinguish the finished track from track, which is typically laid for temporary work, i.e., for transporting material on construction sites.

Desford Colliery Sidings, 1983. There are several items of interest in this photograph, Firstly, note the use of bullhead rail, common in sidings even into the post-Privatisation era. Secondly, note the dilapidated state of some of the sleepers, and how poor the ballasting is. Finally, note the use of a red rag suspended between two sticks on the right, this is presumably a crude method by which to indicate this road is out of use (it's likely any driver would have already deduced this by the presence of substantial weeds)! (*Jamerail/Flickr*)

CIVIL ENGINEERING AND THE PERMANENT WAY • 107

By 1981, the Permanent Way accounted for approximately 40 per cent of the total capital investment in the railway network.

Basic Track Layouts and Nomenclature

Before we design a layout, it makes sense to first familiarise ourselves with the basic nomenclature used by Engineers on the prototype.

The basic 'standard gauge' railway track in use in the UK consists of two parallel rails placed 1435mm apart. The rails are fixed to sleepers which are commonly made from wood but can also be made from cast concrete or steel, the function of which is to support the track and keep it in the correct alignment.

The distance between the two running rails is commonly called the 'four foot', the distance between two tracks is commonly called the 'six foot' and the distance between two pairs of tracks running in opposite directions is commonly called the 'ten foot' by railway staff (these all being approximate Imperial dimensions of the aforementioned distances). The area either side of the railway line immediately adjacent to the rails is known as the 'Cess'. The origin of this word is unclear

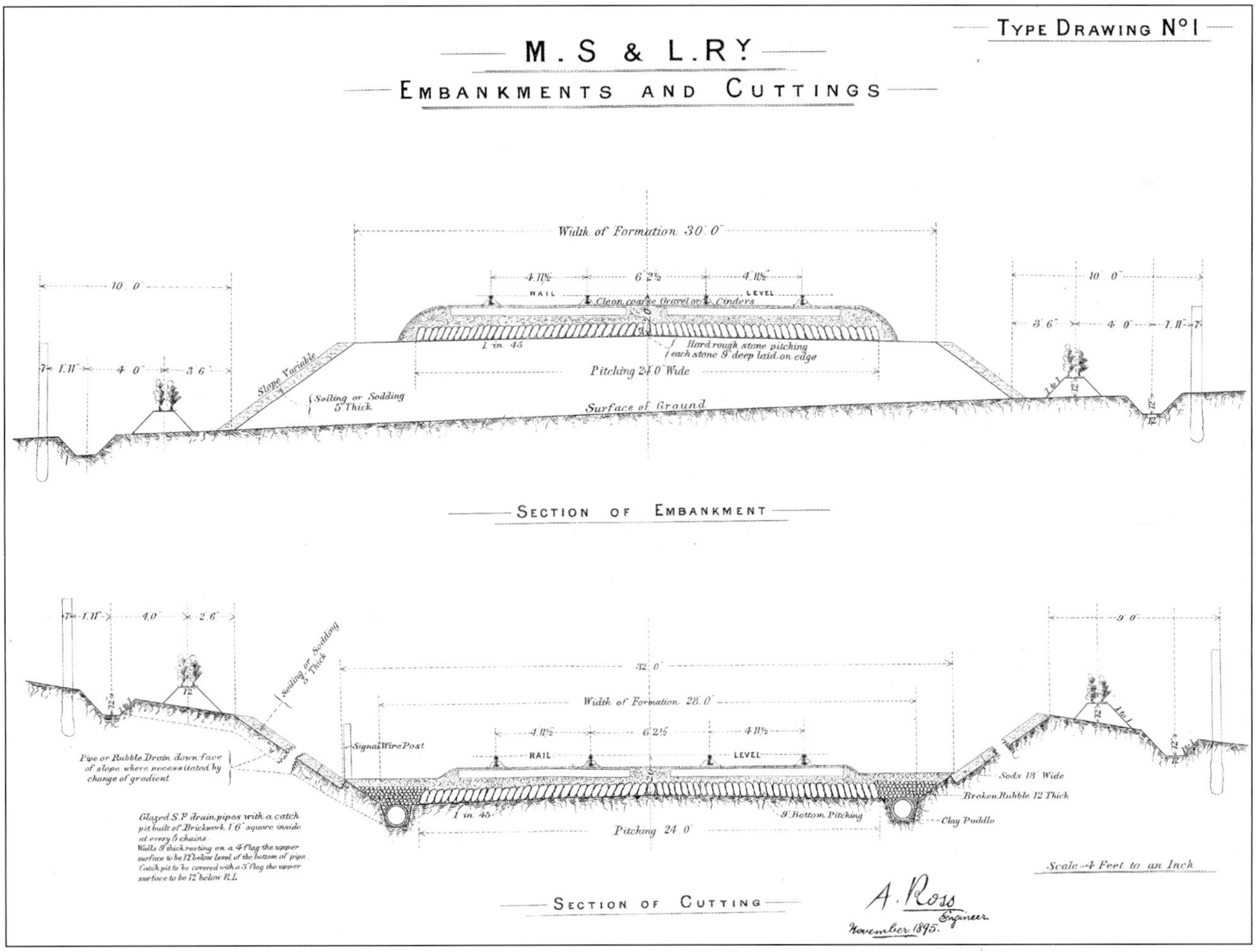

1905 Drawing showing a cross-section of the permanent way on the Manchester, Sheffield & Lincolnshire Railway. The salient features of the permanent way on both an embankment and in a cutting can be observed on the drawing.

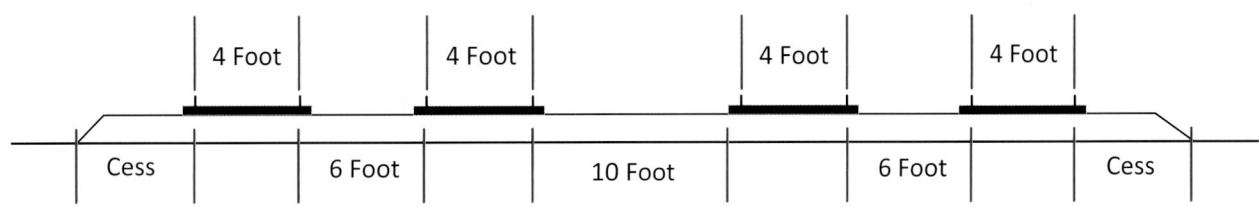

Basic track nomenclature for quadruple track layout.

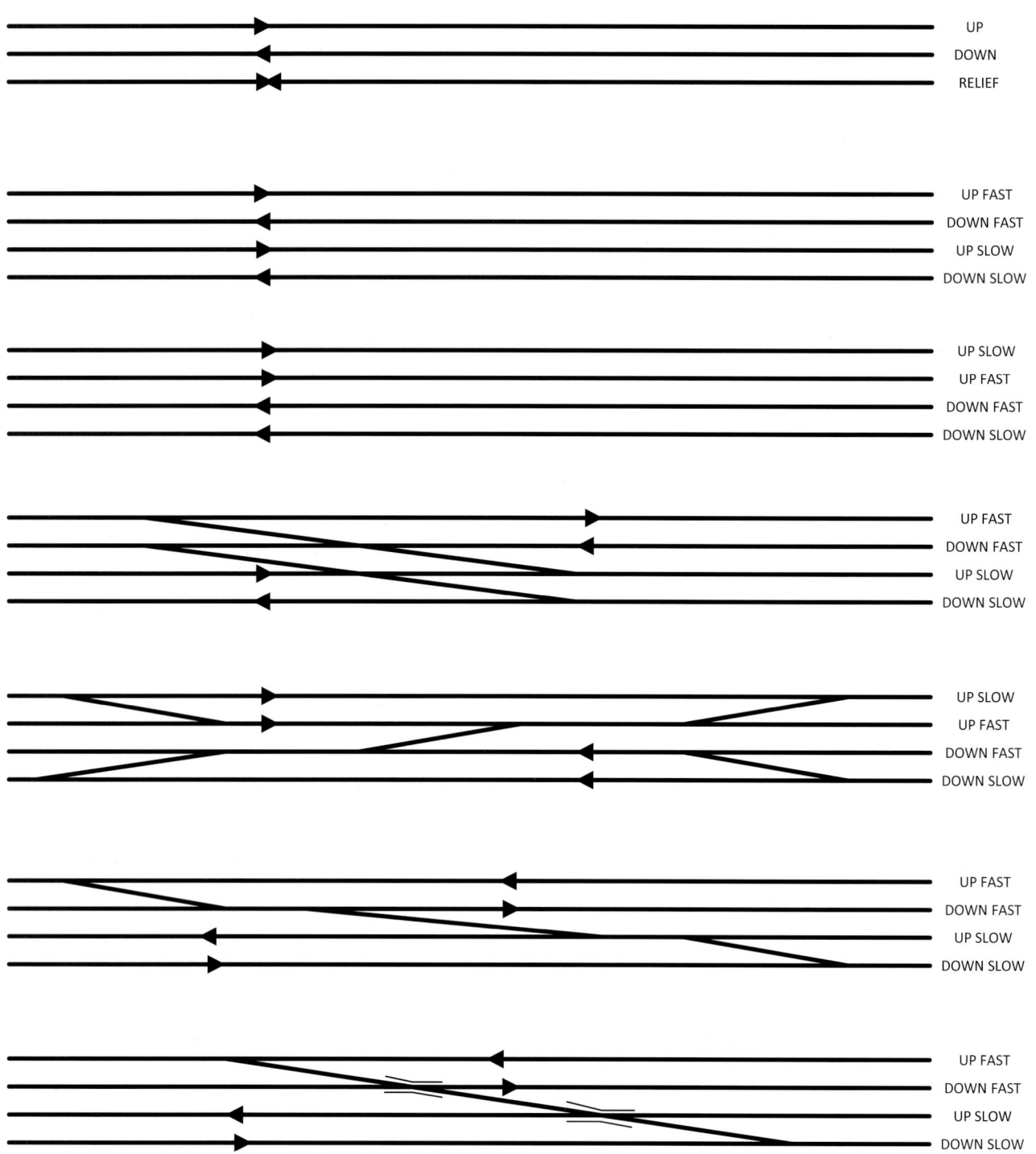

The arrangement of multiple tracks for principal routes.

but a 'cess path' is an old name for a path across a peat bog.

Running lines may be called 'fast', 'slow' or 'goods' and 'Up' or 'Down'. Fast (or main) lines are those used by passenger and express freight trains. Slow lines are also used by passenger and freight trains, typically those operating at lower speeds. Additionally, a 'goods' line was not authorised for passenger trains.

'Up' and 'Down' refer to the direction of the lines. Typically, Down lines are those running towards London whilst Up lines are those running in the opposite direction.

The double track with bi-directional relief road was an uncommon arrangement and was only employed where a single-track branch joined the main route some distance from the junction station. A relief road was sometimes used to enable the branch service to access the station and normally led directly to the correct local platform face.

Quadruple track featured two adjacent pairs of tracks and is the most common configuration for main routes. This arrangement often arose when a second set of tracks was laid parallel to the original route at a later date. This type of layout is also advantageous for the modeller in so much as it allows for the use of a pair of independent electrical circuits. The major issue with quadruple track layouts is their size; most modellers simply don't have the length to fully exploit its advantages.

An evolution of the quadruple track is the quadruple track with outside slow lines. This type of layout permits changeover between fast and slow lines without conflicting track movements.

A double crossover linking parallel tracks allows trains to switch tracks, but in doing so will block the opposing road. This is a visually impressive layout but uses a significant amount of space. This layout became increasingly rare in the diesel age.

Tracks with outside connections for slow lines allow trains to switch from fast to slow lines without interfering with trains running in the opposite direction.

Ladder junctions were favoured during the 1980s. They connect the fast and slow lines using high-speed turnouts with long leads, meaning there was no need for trains to slow to a crawl before negotiating the layout.

Finally, a 'ladder slip' connection saves space and reduced the 'zig-zag' effect of a ladder junction. This configuration has the major advantage of saving space; a boon for the space-starved railway modeller.

Rail

Steel rails were first used in 1857 and by the early twentieth century had become ubiquitous. Rails are typically specified by units of weight per linear length (i.e., 113lb rail is rail that weighs 113lb per foot length). Continuous welded flat-bottomed rail was introduced into Britain in the mid-1960s, and gradually grew in prominence, particular for main lines where high speed and

Flat bottom continuously welded rail (CWR) with concrete sleepers (Top) and bullhead rail on timber sleepers' (Bottom).

smoother running was required. However, for many years, indeed even to this day in some locations, traditional 'jointed' bullhead and new continuously welded flat-bottom track existed side-by-side.

The 'standard' rail type used by the 1970s for new works was 113A. For most new lines, it was laid as CWR apart from in switches crossings, and where the track geometry resulted in tight curves. Bullhead type rails were only really used for new works in special circumstances such as check rails on viaducts. However, there were still large quantities of obsolete rail types utilised all over the network, particulary in yards and on lightly used routes. The main types of rail section used by British Rail in 1981 are listed in the table below.

Types of Rail Used on the BR Network in 1981

Rail Section	Pound Per Yard	Kilogram Per Metre	Use
Flat Bottom			
113A	113	56.1	Main Lines – Switches and Crossings and CWR.
110A	110	54.6	As 113A. Undergoing replacement with 113A.
109	109	54.1	As 113A. Replaced by 110A.
98	98	48.6	Obsolete, used on secondary routes as serviceable CWR
Bullhead			
95	95	47.2	Used as check rails on heavy curves. Previously used on main lines but largely replaced by Flat Bottomed Rail. Still used on secondary and minor lines and in yards, sidings and industrial railways.

Sleepers and Ballast

With the development of effective wood preservatives railway sleepers made from wood became standard. Porous hardwood was used for sleepers as this absorbed more preservative and wooden sleepers had an average service life of over twenty years.

The first general introduction of concrete sleepers on main lines took place in the late 1930s. These were of conventional cast design, direct replacements for the timber type, and were fitted to carry chairs and bull-head rail. Concrete sleepers weigh more than wooden types but had a life in excess of fifty years, making them economically viable for use on certain routes. With the adoption of mechanical handling equipment, the weight became less of a problem and the use of concrete sleepers became more widespread.

By the late 1980s, BR was replacing approximately 1 million sleepers annually, 50 per cent of which were concrete.

Steel sleepers found only limited application on the BR network. The major weakness of steel sleepers is their relative lack of weight, which creates an issue when laying pre-tensioned welded rail, therefore, they were not generally used on main lines.

The track required a stable base to support the weight of trains, keep the correct alignment and provide drainage, this is known as the track bed. The original wagonways were built to carry coal and ore to the ports, the ships arrived filled with a 'ballast' of broken stone and this was used by the early railways to form a path on which the track was laid and the term 'ballast' has remained in use since.

Most lines used crushed stone ballast. In the past, other ballast materials were also used based on geographical location. Welsh lines, for example, used broken slate in the north and the dark waste from lead mining in the south. However, as the years progressed, this practice was largely abandoned and by the BR era, ballasting was almost exclusively undertaken using crushed granite on main routes.

Some ballast contained toxic materials (such as the lead mine waste) which inhibited vegetation growth; where this was not the case it was necessary to treat the track with

Steel sleepered track. (Jamerail/Flickr)

weedkiller to prevent the plants clogging up the ballast and reducing drainage, this being carried out manually, or, for longer stretches, a special train which sprayed weedkiller directly onto the track bed.

In yards and on industrial railways where speeds were slow and uneven track was less of an issue, a simple trackbed of ashes and/or cinders was often laid. Track in such areas was often not built on raised earth banks, this being observable where such lines were laid close to the main line.

Switches and Crossings

Switches and crossings are specialist assemblies used to fulfil the important function of allowing a train to change from one track to another.

Switches and crossings are built up from three principal components: the switch, the common crossing and the obtuse crossing. Switches are normally installed in pairs and handed, i.e., left or right. A variety of sizes are available, during this time period sizes A-F, Short G and G were in common usage. An A switch has a sharp angle of divergence compared to a G switch. Angles are largely derived from planing of the rails in order to achieve the required geometry.

Common crossings originally consisted of two wing rails and a 'vee'. A vee was originally manufactured from a point rail and a 'splice', but increasingly, parts of the crossing were manufactured from manganese or welded and then machined, both types of crossing having greater resistance to wear.

An obtuse crossing is similar in appearance to the common crossing, but its purpose is different. In a similar fashion to common crossings, original 'built up' units were being increasingly replaced by cast manganese crossings or switch diamonds during this era.

Finally, during the late 1970s a new type of crossing was introduced called the 'swing nose' type crossing, primarily for use on high-speed routes.

Crossings are categorised by the angle of intersection between the faces on which the wheel flange sits measured diagonally, for example 1 in 8 (7° 9′ 9″). Built up crossings could be assembled in most geometries between 1 in 4 to 1 in 10¾, with cast and welded crossings only being available in certain sizes.

Combination of switches and crossings are used to create a variety of track geometry. The combination of single leads and crossovers was a feature of new track installations during this era, and permitted increased speeds through complex trackwork, improving efficiency.

Most modellers will be constrained by the proprietary track system they use but the rule of thumb is to use the longest point with the shallowest crossing angle you can fit in the space available. This is especially important where an 's' curve is involved, such as when a train has to cross from one line to the other on a section of double track. Failure to follow this rule can result in derailments.

The most common combinations of switches and crossings used on the railway are listed below:

- Left-Hand or Right-Hand Turnout – This enables the train to diverge left or right (depending on the configuration of the turnout). If two turnouts are placed together a crossover is created.
- Contra-Flexture Turnout – This enables a secondary line to curve away at a sharper angle to the main line.
- Three-Throw Turnout – A Complicated installation only really used in areas with restricted space such as the entrance to stations or yards. Enables a train to go one of three directions.
- Diamond Crossing – Allows two lines to cross one another. Does not allow a train to move between the tracks.

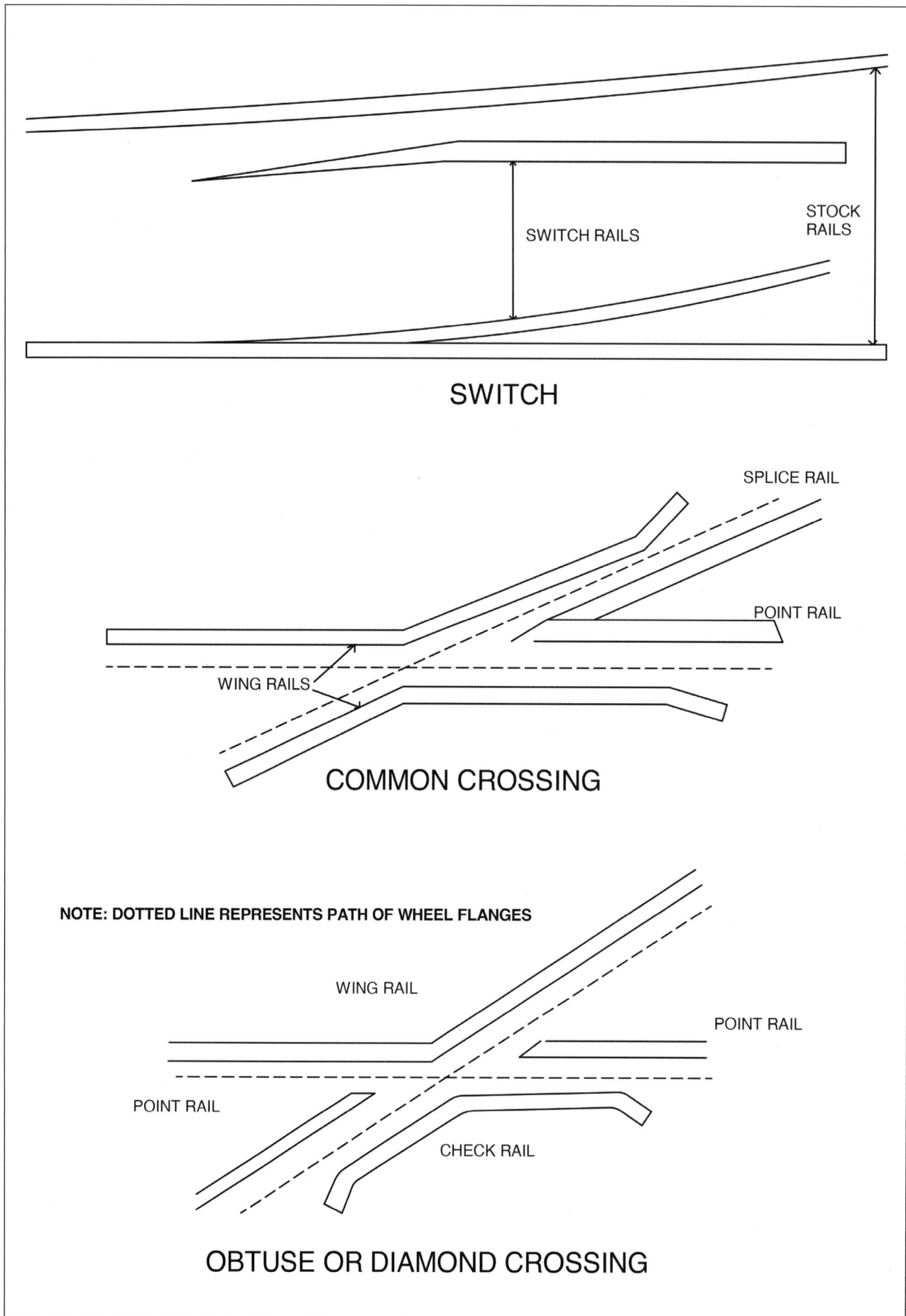

Common crossings.

Types of Single-Lead Crossover in Use on BR as of 1981

Type of Switch	Crossing Angle (1 in ___)	Radius of Turnout Curve (m)	Maximum Permitted Speed (MPH)	
			Single Lead	Crossover
Av	7	141	20	15
Av	9.25	*	20	-
Bv	8	184	20	20
Bv	10.75	*	20	20
Cv	9.25	246	25	20
Cv	13	*	25	20
Dv	10.75	332	30	-
Dv	15	*	30	30
Ev	15	645	40	-
Ev	21	*	40	40
Fv	18.5	981	50	-
Fv	28	*	50	50
SGv	21	1264	60	-
SGv	28	*	60	60
Gv	24	1650	70	-
Gv	28	*	70	70

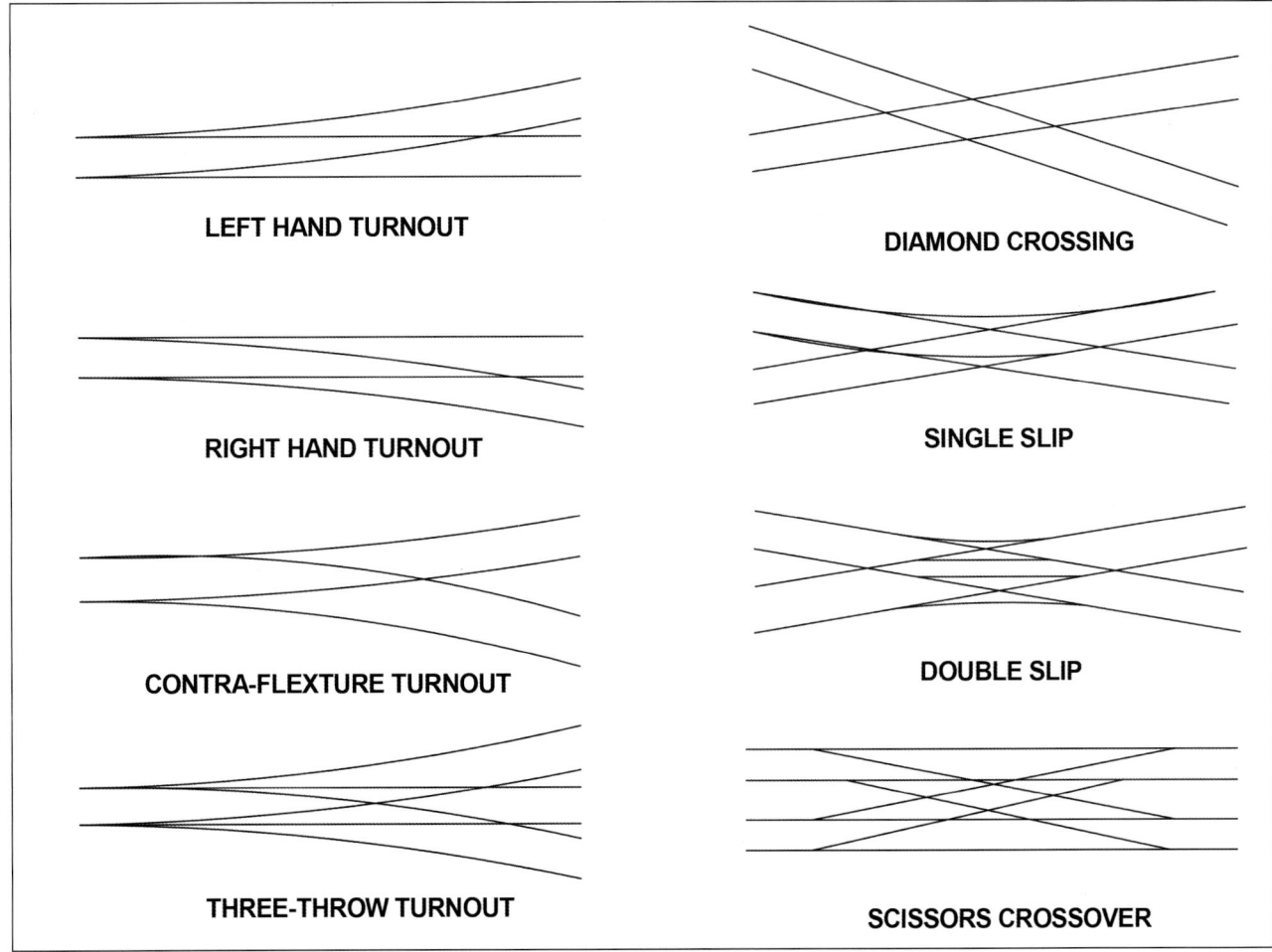

Common combinations of switches and crossings.

- Single Slip – Allows trains one of two lines to cross the other line or to change route. Does not allow a train on the other route to change track (i.e., it can only cross).
- Double Slip – Allows trains on either track to both cross the other line and change track.
- Scissors Crossover – Trains from both directions can change track. Normally only seen within stations and increasingly rare in the diesel era due to the cost.

Turnouts on running lines used by passenger stock were arranged wherever possible in a trailing direction, that simply means that the train would normally never encounter a diverging track on a running line. Trains were normally required to reverse over points on running lines to change from one line to the other. Where this could not be arranged, the point was fitted with a locking device. This could be a lever operated device connected to the signal box (in which case the point would have a protective bridge in wood or metal over the mechanism). In some cases, the point was manually locked for main line running using a padlock.

Catch and Trap Points

After an accident in the 1870s where wagons rolled back out of a goods yard and fouled the points on the main line, a new regulation was introduced requiring the installation of special pointwork to prevent runaway vehicles fouling the mainline. Such points resemble a conventional set of points, but they did not connect to another track, their function being to derail vehicles rolling from a yard or siding toward a running line. There were several configurations, not all of them still in use by the BR era, and they were referred to under the umbrella term 'safety points'.

For some reason, probably because they are difficult to model unless using hand-built track, and the fact that they don't actually do anything on the model, safety points are omitted from most layouts. However, if modelling certain types of layouts, the inclusion of safety points (or at least a representation of them) can do much to add authenticity to the scene.

The basic difference between a catch and a trap point is that a trap point is designed to derail stock that runs too far forwards; whilst a

A Trap Point can be seen in this 1976 photograph of Doncaster Station. The trap point is protecting the passenger lines ahead of the line third from the right. (*David Flitcroft*)

catch point is designed to derail stock that runs backwards.

A trap point was normally found at junctions between sidings/secondary lines and the main line. The installation basically consisted of a point that would be 'set' to derail anything leaving a goods loops or siding. Often the movement of the point was interlocked with the signalling to prevent accidental operation.

Deliberate derailing of runaway vehicles was considered a last resort and often resulted in damage to vehicles, and they then needed to be manoeuvred back onto the rails to clear the line for other traffic. Some locations used a full point, with a large sand trap to catch runaway vehicles. This was called a 'drag'.

A catch point was installed at the foot of an incline, the fitment of such devices mandatory on gradients greater than 1 in 260. Catch points resembled trap points but were slightly different in the fact that the point blades were spring loaded. A train going up the hill would pass through the point but any loose wagons rolling back on the wrong line would be derailed. This was particularly important in the case of unfitted wagons as they had no automatic brakes and once detached from a train could easily run away.

Point Motors, Levers and Groundframes

By the 1960s, point motors were in widespread use and although a number of designs were utilised, they all allowed the remote control of a turnout or crossing. Where manual control of pointwork was required (such as in yards or sidings), a manual lever would be provided. If a number of levers were grouped together, these were incorporated into a device known as a groundframe. Groundframes often incorporated a platform upon which groundstaff could stand; the platform would often have the name of the groundframe fixed to it in a prominent location.

Point heaters would often be located in proximity to point motors, sometimes heated electrically, but bottled gas was also used, particularly in locations away from urban areas. The location of a heater would be evident by the presence of a cage containing one or more gas bottles.

Electrically heated point motors seen here at Newcastle Station in 1987. (*David Ford*)

Expansion Joints

During service, steel rails would expand and contract depending on the temperature. With traditional bolted track, an allowance was made within each joint which took into account this expansion and contraction, without unduly impacting the alignment or geometry of the track.

Continuously welded rail could not readily incorporate these gaps as there were no joints where a gap could be provided. Therefore special 'expansion joints' were incorporated into such layouts. As the lengths of rail involved are long, the cumulative expansion is large so at the joint the ends of welded rail are cut to wedge shape and a special chair is fitted to hold the ends together.

The joints themselves are approximately 450mm long; the primary visual identifier is two lengths of bullhead rail bolted across four sleepers bridging the joint. The rail is approximately two-thirds the height of the running rail to ensure no contact with passing vehicles.

Other Permanent Way Features

There are, of course, a number of distinctive features that are part of the permanent way but are not specifically part of the track. In this section we will describe and illustrate some of the more common features.

One of the most common and immediately obvious items often included on layouts and on the real railway are buffer stops. These are constructed from a variety of materials and serve to arrest the movement of a train and prevent it from travelling beyond the end of a line. Buffer stops were not intended to stop a runaway train, although they might arrest the movement of a wagon at the end of a siding. When positioned in platform bays or at the end of the main line they carried a lantern showing a white (later red) light.

The most basic configuration was a timber beam, frequently an old sleeper, laid across the track, with a warning flag and/or lamp placed to make it conspicuous. Such temporary stops were frowned upon on the

Expansion joint.

An example of an improvised buffer stop consisting of an old sleeper with red flag and tail lamp to warn drivers that the track doesn't reach the end of the platform bay. (Jamerail/Flickr)

mainline railway, but they were sometimes used on out of use lines/sidings, in stations, and on industrial railways. Use on the main network was normally restricted to temporary installations.

A cruder option, sometimes seen on industrial lines and private railways, was to bend the ends of the rails upwards a short distance to act as a buffer stop. This allowed the wheels of the rolling stock to act as its own buffers. The rail was bent high enough to be above the centreline of the wheel, and thus the wheel was unable to roll-over it.

A step up in complexity was the 'sleeper built' buffer stop which was a box built from redundant timber sleepers set into the ground and filled with earth or sand. This type of arrangement appears to have been relatively uncommon.

Probably the most common type was the rail-built buffer stop. This consisted of lengths of rail bent to an inverted U shape and bolted to the rails with the same sort of bolts used for fishplates. To these were bolted vertical lengths of rail, extending up from the track. At the top of these vertical rails there would be either a timber sleeper or a couple of lengths of rail bolted to the frames at the same height as the buffers on rolling stock.

A less common variant was the timber-built buffer stop, consisting of two heavy vertical timbers set into the ground with a similar beam mounted horizontally at buffer height and behind this were two more heavy beams set into the ground at an angle to brace the frame. The vertical posts typically extended some distance above the top of the buffer beam. These would require rather more work to maintain than the rail-built type (which could be unbolted from the track and removed or replaced entirely) and could only be used at the end of a siding where the track ended (the rail-built type could be bolted on before the end of the track, so the rails did not need to be cut to length). In addition, 'scrap' rail was freely available and cost virtually nothing.

At the ends of some sidings, the track was sometimes banked up, so that a few feet were on a rising gradient. This served two purposes; firstly, it would help slow down a rolling

An example of the rising gradient buffer stop seen here at Longsight Depot. In BR days, the LED light would have been an oil or battery tail lamp.

vehicle before it hit the buffers and secondly, when the siding was being shunted, the engine driver would see the end vehicle rising as it approached the end of the line and thus had a reasonable indication as to the position of his train. Examples of this can be found in yards and sidings throughout the network, although it doesn't seem to have been a widespread practice (probably due to the effort required to create the required earthworks).

Buffer stops were often painted red or white (occasionally with a black or white band in the centre of the stop). From the 1990s onward, some buffer stops were also painted yellow. However, more generally a 'rust' colour would be a better representation of the metal type unless newly installed as they appeared to receive scant maintenance attention.

In main line terminus stations, there were experiments with hydraulic buffers. They consisted of a large concrete block with a pair of very long shank buffers sticking out of them. The use of such installations was limited only to principal stations, probably due to the cost and maintenance burden associated with a buffer stop of this type.

There were also all manner of specialist equipment fitted to locations throughout the network including Hot Axle Box Detectors (HABD), flange greasers, lineside relay boxes, rodding for pointwork and so on. Many of these items would be very small in N Gauge, but close observation of photographs will reveal a surprising amount of 'stuff' on a length of track, and the representation of this equipment is something rarely seen on layouts.

On the Model – Operation and Appearance

The provision of realistic trackwork in N gauge, has, until fairly recently, meant scratchbuilding. However, there are now two commercially marketed proprietary track brands that offer something approaching prototypical accuracy,

Both brands offer a range of realistic turnouts and crossings, with either bullhead or flat bottom rail and timber or concrete sleepers. Assembly is required, although the preformed bases, precast crossings and available jigs do make the task less arduous. However, there is no escaping the fact that creating a substantial layout using Finescale track takes a lot of effort, effort many modellers would rather put into the creation of the scenery and the trains. Thus, the majority of modellers will utilise pre-assembled track from one of the commercial manufacturers. Despite the incorrect scale/gauge ratio, there is still much that can be done to commercially available track to improve its appearance.

For realistic operation, trains will need to run smoothly without derailing. On shunting layouts, slow speed running will be essential. Therefore, it is vital that track is properly laid, clean and properly tested. It is also very important that all clearances are checked using the widest piece of stock prior to being permanently fixed to the baseboard.

In addition to ensuring the track is laid and wired properly and ensuring that it is properly cleaned for good electrical conductivity, we should also consider the appearance of the track. Next to locomotives and rolling stock, track is one of the most noticeable visual aspects of a model railway. Whenever a model railway is viewed or photographed, it is likely some track will be in view.

The sleeper dimensions and spacings of Peco track is incorrect for UK layouts, as the track was designed for American N Scale. Sleeper spacing can be improved by snipping the web between the sleepers and removing alternate sleepers to represent UK track more accurately.

With all trackwork, the sleepers have an obvious plastic sheen, and the rails are a shiny nickel-silver. Therefore, a simple and instant improvement can be gained by painting and weathering the track. The colour of the track varied based on the age of it, how much traffic it saw, the state of repair and the location. The rails and sleepers usually appeared a dull brown (apart from concrete sleepers) and the tops of

This photograph taken at Ipswich shows to good effect the colour differences on the track where diesel locomotives and multiple units stand. Note the black gunge and sheen between the rails, grease and oil from standing trains, and the dirty but differently coloured ballast in the six-foot. (*Jamerail/Flickr*)

This picture, taken at Liverpool Street in 1983, shows the atrocious state of the track and ballast even in important stations which was a 'feature' of the run-down network in the 1980s. Note the rubbish, toilet paper and thick gungy grease coating the rails, sleepers and ballast. (*Jamerail/Flickr*)

the rails for any track that received trains would be shiny. This can be used to good effect to show a long abandoned or disused section of track, i.e., simply painting the entire track including the rail head the same grimy, rusty colour.

There were often black stains from oil and grease along the centre of the track. This was more pronounced in areas where trains stood, such as depots and sidings, in some cases the entire four-foot will be black. In addition, there was often grease around points, which migrated to the surrounding ballast and sleepers. There was also often spillage or dust blown from passing freight trains, black coal dust for example. Another rather unpleasant, though visually distinctive, feature of the track in this era was the presence of effluence and toilet paper on the track from passing trains; used toilet paper in particular would stand out against the dark colours of the ballast.

Reference to photographs of the prototype will do much to help the modeller paint and weather track to a realistic shade.

Railway Electrification

A railway electrification system supplied electrical energy to railway vehicles so that they can operate without having an on-board prime mover. Electrification requires significant capital expenditure and this has meant that in the UK, its application has been restricted, primarily to heavily urbanised and principal routes.

The major electrification event before nationalisation in 1946 was the extension of the Southern Railways system over the main lines from London to Brighton, Eastbourne and Portsmouth, and along the coast towards Ore. This, coupled with a rapid conversion of suburban routes such as Reading and Gillingham, quickly created a large, electrified network in the south. This system was based on 750v DC and used conductor rails. Elsewhere in the country, the pattern for electrification was patchy at best.

In the years immediately following the Second World War, experiments were undertaken utilising industrial frequencies as a traction supply. Early hopes of developing a 50Hz AC traction motor that could compete with the existing DC motors failed, but rectifiers suitable for installation in rolling stock had by this time become available and so the solution adopted was a high voltage supply at 50Hz to the contact wire, which was then transformed down and rectified onboard the vehicle to provide DC at a suitable voltage for the traction motors.

The modernisation plan of 1955 included proposals for further electrification of main lines and suburban routes. In early 1956, BR announced all new electrification works outside of the Southern Region would be equipped with a 50Hz system operating at 25kV.

By 1979, the total electrified routes exceeded 2,300 route miles, 1,125 route miles being high-voltage AC and 1,185 miles being 750v DC.

25kV AC Overhead Line Installations – An Overview

The standard BR system takes power from the National Grid, usually at 132kV, at feeder stations spaced 20–30 miles apart. In the grid stations, the voltage is stepped down through a series of transformers to 25kV and then fed to the trackside feeder stations and finally into the trackside catenary through circuit-breakers. Because the National Grid is a three-phase system and the railway system is single-phase, the phase connections at each feeder station are chosen as to minimise unbalance of the three-phase system. Supplies from the grid must not be interconnected, so there is a break in the electrical continuity of the railway overhead system by a neutral section half-way between feeder stations. There are usually similar neutral sections at each feeder station, where as far as possible supplies are taken from two different sources which are independent of one another. Electrical switchgear is provided enabling neutral sections to be bridged through busbars fitted to the lineside equipment. This would typically occur if the interruption of the electrical supply makes it necessary to change the normal pattern of feeding.

The 'dead' section of contact wire within a neutral section is separated from the 'live' sections either side of it via the use of insulators which have wires of a slightly larger cross-section than the wire along which the pantograph of a vehicle slides. The length of a dead section is typically only 5m. At neutral sections, magnets are fitted in a similar manner to the Automatic Warning System (AWS). They open a circuit breaker within the vehicle as it passes over the magnet, which ensures that the vehicle does not take power when travelling over the neutral section. A magnet at the other end of the section does the reverse. This system is known as Automatic Power Control and is designed to prevent damage to vehicles and the overhead line equipment.

Another type of insulator used where there are switches and crossings is known as the section insulator. Whilst this separates adjacent sections of contact wire, it does not interrupt power to a vehicle pantograph, being arranged so that the pantograph of a passing vehicle is continuously in contact with one or the other section energised wire.

The catenary system from which trains collect power is erected in lengths of up to a mile. Each full length is restrained at one or more structures in the middle and tensioned via adjustable weights at each end. Adjacent lengths overlap so that power to passing vehicles is not interrupted.

In the early stages of electrification, a number of systems were in use, but by the mid-1960s BR had developed a catenary system known as the MK III A. A similar system, the MK III B was used during the St Pancras–Moorgate–Bedford electrification scheme in 1976.

The MK III A and MK III B systems are relatively simple systems, with the contact wire suspended via droppers from a single conductor. The span of a MK III A system was a maximum of 73m and there was a sag of 100mm in the contact wire at the half distance point of the span. The purpose of the sag was to improve current collection at high speeds by presenting a level contact wire to an uplifted pantograph. The wire lifts as the train passed, and the lift is greater at mid-span than at the supports. This lift is compensated for by the sag in the wire.

On a single or a double track, the catenary is supported by individual masts. The contact wire does not follow the line of the track, but 'zig-zags' from one side of the span to the other,

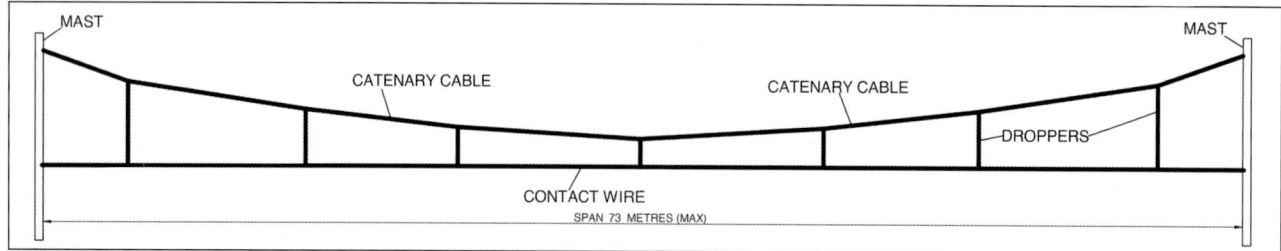

Catenary nomenclature.

in order to ensure even distribution of contact wire wear.

On quadruple track, the catenary is supported by a construction known as a headspan with the exception of the restraining structures at the centre of the span, the catenary cable is carried in pulleys and the contact wire is located via a steadying arm. The cross-span wires directly below the headspan are copper-clad in order to carry any fault current.

Historically, if an N gauge modeller wanted overhead catenary this meant scratch building or constructing etched kits. However, pre-built catenary from Dapol is now available, which simplifies things somewhat, although the range is limited. There are also extensive catenary ranges in European catalogues, but these do not resemble UK catenary and require modification in order to achieve an acceptable likeness. The 'knitting', i.e., the wires between masts, is something that only the most dedicated modellers will want to attempt, the wires have to be sufficiently fine to achieve a reasonable appearance and this will result in a fragile installation immediately above the running track, which can make cleaning and maintenance challenging. However visually stunning a complete OLE system looks, it's generally accepted that most modellers wishing to represent the catenary will be content with just the masts, insulators and support tubes.

750v DC Third Rail Installations

Lines operating on the 750v DC third-rail system take three-phase AC power from the national grid and distribute it via lineside cables to

In this shot taken outside Birkenhead North EMU depot in 1983, conductor rail and wooden guards can be clearly seen. (*Jamerail/Flickr*)

substations where the supply is stepped down in transformers to the required voltage.

Distances between substations vary between under a mile for very busy areas to 5 miles in rural areas. The high voltage cabling between the track and the substation is of a thick oil-filled type, which enables it to work at high temperature and carry a high load.

The 'live' conductor rail is mounted on insulated pots outside the running rails. To complete the circuit the traction current flows through the wheels and axles of a passing train, into the rails and back to the substation.

Midway between substations there are usually gaps in the third rail which are bridged electrically via a busbar located in a lineside hut. Circuit breakers located in the same hut open automatically to cut off supply if there's a fault and can be manually operated if the section of line is required to be manually isolated in the event of an emergency or for maintenance.

Current is collected from the conductor rail via a conductor shoe located on the bogie of the rail vehicle. The shoes are sprung-loaded via a resilient bush to provide a constant contact pressure between the shoe and the rail.

The third-rail system is rather more straightforward to represent in N Gauge than overhead catenary. Code 40 rail can be fixed to a plastic base and laid alongside proprietary Code 55 or Code 80 track. The use of Code 40 rail has a couple of advantages, firstly it is easier to bend at the end of the section, being of a thinner cross section, and secondly, as it is below the height of the running rails, fouling of passing trains is less likely. An approximate gap of 2mm between the running rail and the conductor rail gives a reasonably pleasing representation of the arrangement.

Civil Engineering

When the canals were first constructed, civil engineers of the day followed the contours of the landscape wherever possible to avoid the need for constructing locks. This in many cases added considerable distances to the route.

When the railways followed in the early 1830s, they wanted to take a more direct route. This meant constructing earthworks, which involved the movement of huge quantities

This atmospheric photograph, taken at Diggle in 1986, shows to good effect the visual aspect of the railway literally cutting through the landscape. A visually accurate layout will have realistic earthworks and civil engineering structures complementing well laid permanent way with prototypically accurate details. (*David Flitcroft*)

of earth, most of which was moved by hand. There are also dramatic cuttings and tunnels cut through solid rock. Invariably this involved manual work with pickaxes, although quantities of blasting explosives were also used.

Properly constructed earthworks make your railway look like part of the landscape rather than track simply laid on a board. They are relatively straightforward to reproduce on a model, but, as with most other things, careful observance of the prototype will yield the best results.

Typically, earthworks can be broken down into two distinct groups, embankments, and cuttings. Embankments are areas where the trackbed has been raised above the surrounding ground. There are many reasons why this was done, but most often it was to provide a stable, well drained bed for the permanent way, unaffected by flooding and subsidence.

Properly executed, such an embankment makes for a striking feature on a layout. Great care had to be taken when building embankments, as the feature itself was load-bearing and manmade, their slopes had to be relatively gentle. The soil used to build the embankment most often came from a nearby location, preferably from a recent excavation on the same railway. When the Great Central Railway was constructed, soil from a nearby cutting was often used to build an embankment further down the line.

The stability of embankments is affected by water, but drainage is relatively easy to manage. Embankments are most at risk from major floods, and, in extreme cases, many have been swept away.

The sketch on the next page shows the features of a typical embankment for a double track route. It is unlikely a drainage ditch would still feature by the 1970s, in most cases, the original ditch would be replaced with a piped drainage system built from brick or pre-cast concrete.

It is possible, on a model railway, to have slightly steeper embankment slopes than the prototype, to save space, but the effect should not be overdone. The typical slope given in the diagram is for soil. For other types of ground, such as clay or rock, different angle slopes would have been used.

In this photograph, we can clearly see the visual impact of an embankment. This is more realistic than ground level track which so often tends to take on the appearance of track laid on a board, rather than an accurate portrayal of a real railway built through an undulating landscape. (*David Ford*)

CIVIL ENGINEERING AND THE PERMANENT WAY • 125

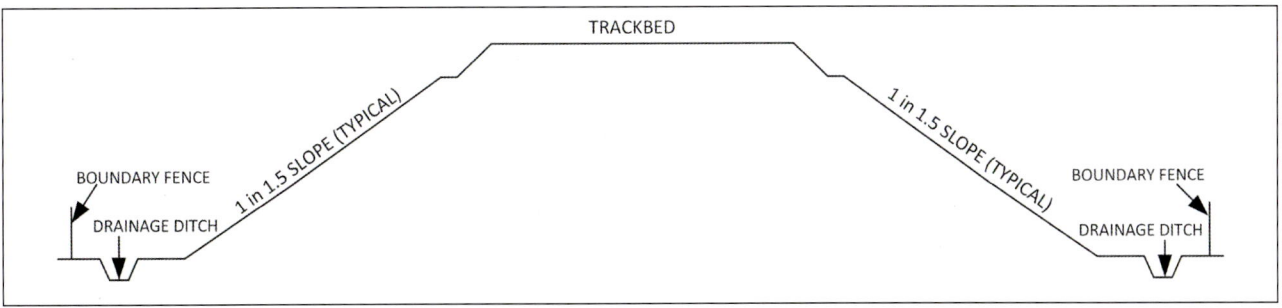

Typical Embankment.

The superb 1980s era layout 'Shirebrook' features convincingly executed civil engineering which complements the perfect weathering and formation of the rolling stock. A Class 25 rumbles across Sheepwash viaduct with High Marnham to Toton tanks. (*Duncan Hunnisett*)

Cuttings, as the name suggests, are channels cut into the ground, usually bedrock, to allow the line to pass through a hill or other feature. They are mostly used in conjunction with tunnels, although short cuttings to allow the passage of a line through rock are also common.

Water is a major problem in a cutting, trapped in cracks it freezes and expands, causing landslides. Precautions must therefore be taken to ensure adequate drainage by providing a drainage ditch at the top of the cutting. If too much water gets to the face of a cutting that has been cut into soil, it can cause it to 'slump' downwards towards the track, potentially blocking the line and putting trains at risk. Often piles of rock or concrete were placed at the foot of a cutting cut into soil to help prevent this movement. Rock filled vertical or 'Y' shaped drainage channels running down the sides of the cutting also helped channel water down away from the cutting face.

Finally, at the base of the cutting, drainage is also required to remove water at track level. When cuttings were originally constructed,

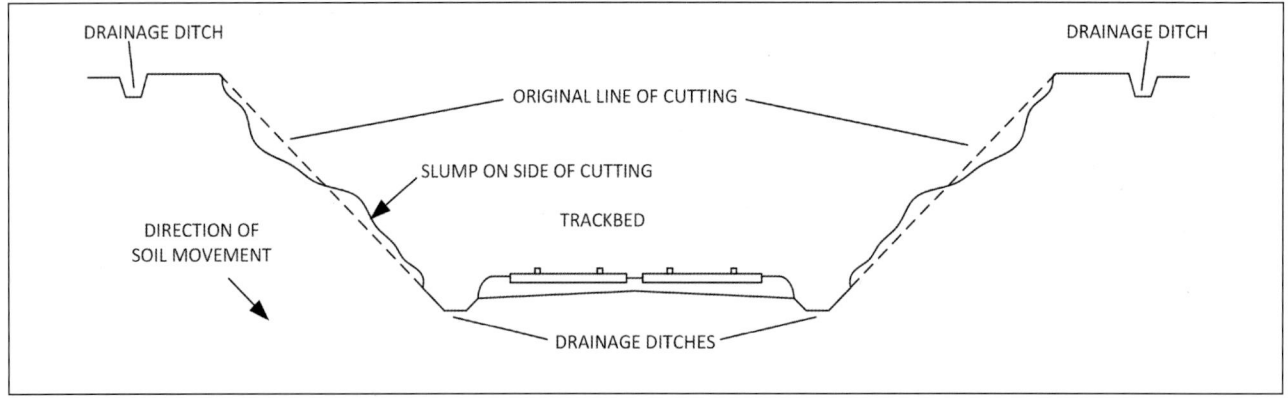

Cutting – typical constructional details.

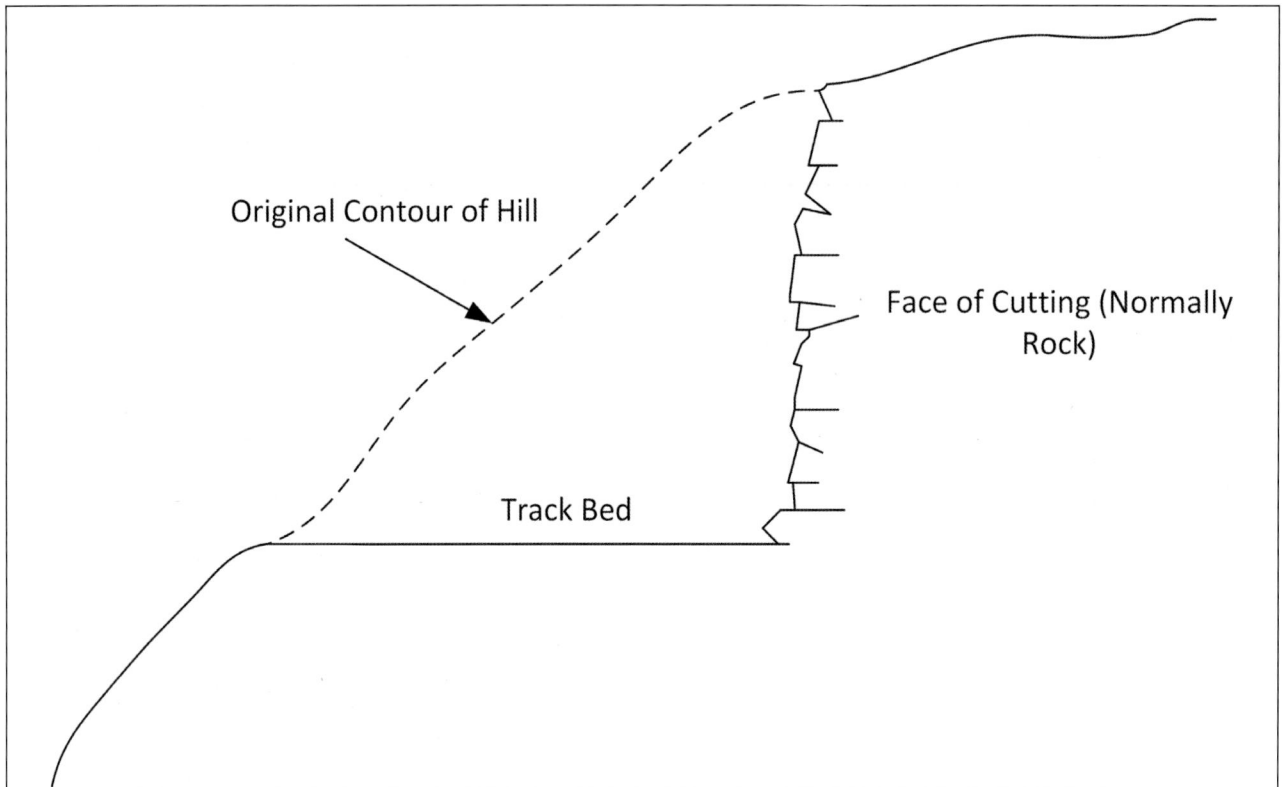

Cutting inside of hill.

simple drainage ditches were dug. In more modern times, it is more usual to see piped drainage used in this area.

Bridges, Viaducts and Tunnels

Bridges

Bridges are one of the most common features on railways. Not only do they carry the rails over valleys and rivers, but they also convey roads and paths under and over the railway.

There are three basic types of railway bridge, masonry, girder and concrete. There are, of course, many forms of bridge within these main groups, and materials were sometimes mixed on the same bridge (girders with concrete for example). Railway bridges last a long time. Awsworth Bridge at Bennerley near Ilkeston in Derbyshire was constructed by the LMS in 1900 and despite many minor mishaps and two serious derailments which damaged the structure of the bridge, it was not replaced until 2016, lasting an impressive 116 years. Therefore, it is entirely possible to have bridges of varying ages on a layout, illustrating the development of the line over many years. It is worth noting that timber bridges were almost extinct in the diesel era and are only really suitable in most cases for period layouts.

CIVIL ENGINEERING AND THE PERMANENT WAY • 127

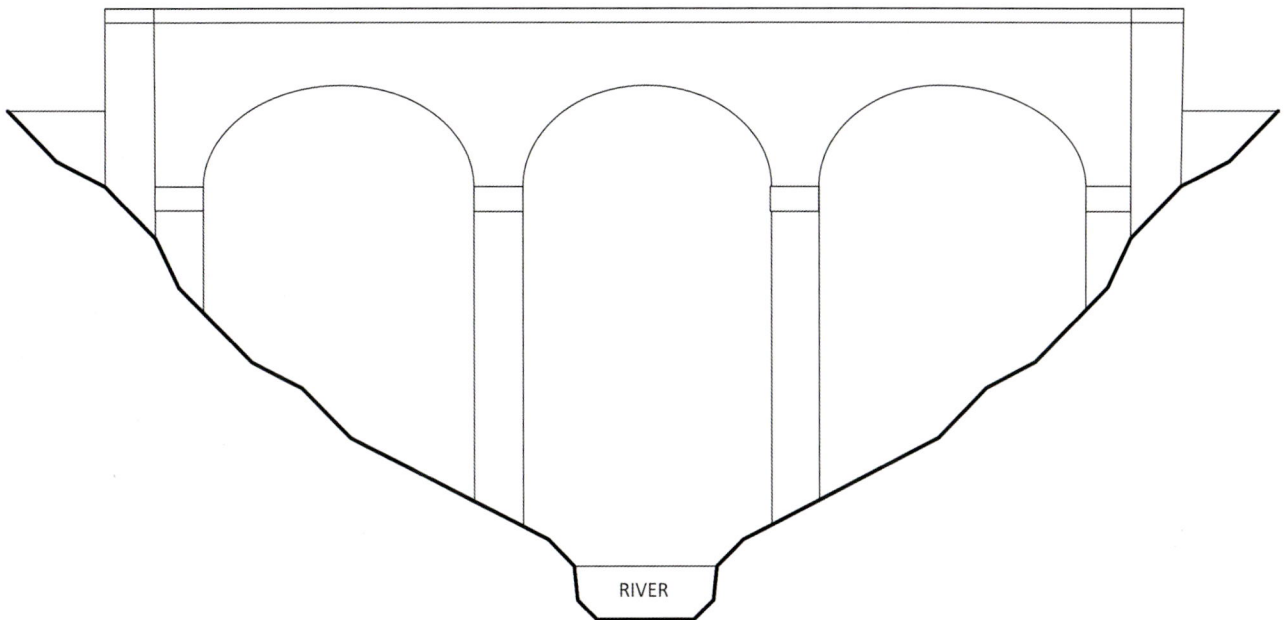

Basic three arch masonry over bridge which would carry the railway over a gorge or deep cutting.

A bridge on your layout can be an imposing focal point in its own right. The Loch Crearan Bridge on the Ballachulish Branch, pictured here in 1974, would make an impressive centre piece to any layout. It mixes the typical brace girder construction with mock battlement stonework at either end. (*John Ford*)

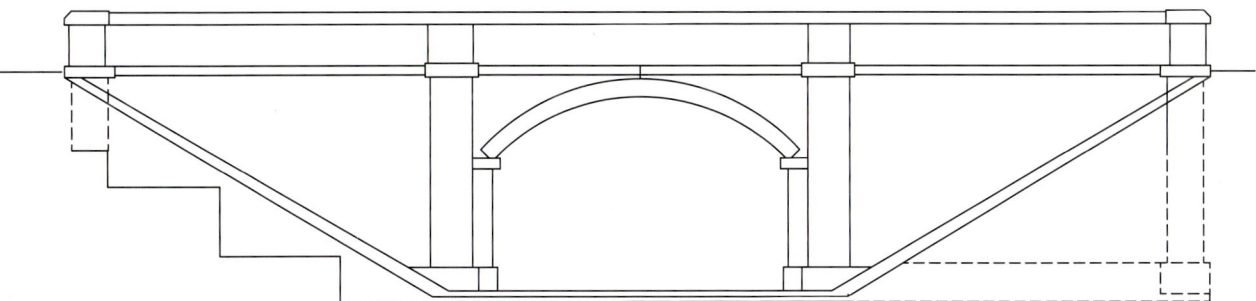

Standard ordinary bridge over or under a railway.

Viaducts

Viaducts allow the railway to cross obstacles in much the same manner as a bridge. Indeed, there is some confusion around when a long bridge becomes a viaduct and vice versa. One common rule of thumb was that a bridge was three spans or less; anything over was generally

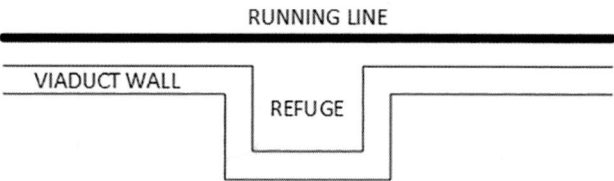

Long bridges and viaducts normally incorporated refuges into the side walls. These refuges allowed track workers to stand in a position of safety as a train went past.

Ribblehead Viaduct, on the Settle-Carlisle railway, seen here in 1972, is perhaps the most famous example of the classic brick arch viaduct. (*David Flitcroft*)

The Great Central Railway Viaduct (front) and Great Northern Railway Viaduct (Rear) in Nottingham, viewed from the Broad Marsh Carpark in 1975. The GCR viaduct incorporates commercial premises between the arches, a common sight in heavily urbanised areas. Note that the GNR viaduct is on a distinct curve. (*John Ford*)

termed a viaduct. Another school of thought was that anything that crossed a valley was a viaduct, regardless of length.

Viaducts were commonly built from brick. However, there were several viaducts built from wrought iron. These were usually required where the land could not sustain the immense weight of a brick viaduct such as areas prone to flooding or with a risk of subsidence.

Tunnels

The purpose of a tunnel on the real railway is to provide passage through an otherwise impassable area. This is normally due to excessive gradients, necessitating excavating through solid rock. During the construction of a railway, a tunnel is constructed only as a last resort due to the huge expense involved. Nonetheless, tunnels are a common feature of many railways.

Tunnels were originally dug by teams of labourers known as navvies. This was hard and dangerous work, and collapses and cave-ins were commonplace. The construction of a tunnel was normally started from a shaft bored in the hillside above, only very short tunnels would be dug directly into the side of the hill. Tunnels were provided with shafts for ventilation. Most often, the air shaft would be the original shaft bored during the construction of the tunnel. The entrance and exit of a tunnel was known as the portal or the adit.

The north portal of the former 'Up' tunnel at Kelmarsh in Northamptonshire. Seen here in modern times, these tunnels were single tracked and derogatively known as 'rat holes' by drivers. The portals and their faces are stoutly constructed from blue brick.

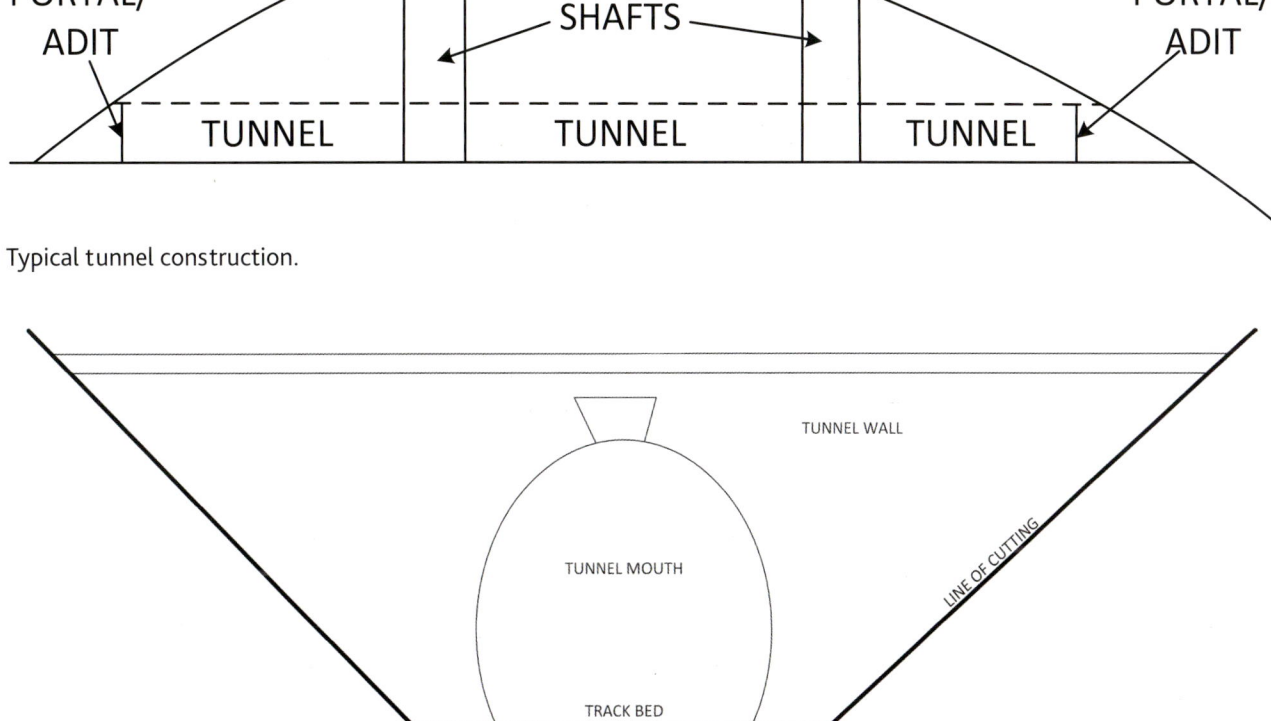

Typical tunnel construction.

Tunnels are normally preceded by a cutting. Elliptical tunnel bores were often favoured, but for maintenance purposes or to combat collapses or flooding, often the bore would be reduced to the vehicle loading gauge in subsequent years.

Tunnel vents are a visual reminder of the presence of a tunnel yet are rarely seen on layouts. This particular example is a vent for Glaston Tunnel in Rutland. Note the wire mesh, provided to prevent items falling into the vent and thus the tunnel below.

This shot of Kilsby Tunnel, taken in 1975, shows to good effect the type of cutting and earthworks commonly associated with railway tunnels. If it were not for the presence of wires, this could conceivably be anywhere in the UK and would make a good model. (*John Ford*)

On the model, a tunnel by itself is rarely modelled as a layout focus, instead they more commonly provide a scenic break at one or both ends of a layout. Occasionally, a tunnel may be included in the middle of a layout as a feature in its own right, but this requires a fairly large layout to be executed convincingly.

A number of tunnel portals are available from manufacturers. However, in order to convincingly portray a tunnel interior, a representation of the tunnel walls and lining should be incorporated. The easiest manner in which to achieve this is a sheet of card formed to the same profile as the tunnel mouth. It would be possible to cover the card in adhesive brick paper, but it's also possible to just paint the card black, unless a light is shone directly into the tunnel itself, the casual observer would not notice that the tunnel roof was not made from brick.

Level Crossings

The traditional railway crossing was manually controlled by a crossing keeper and fitted with wooden gates. This type of crossing survived in some locations well into the 1990s, although in later years the crossing was often only manned on a part-time or ad-hoc basis.

Automatic barrier crossings were provided with full width barriers that prevented passage across the crossing once a train is on the approach. Such crossings were normally also fitted with warning lights and a bell.

Automatic half barrier crossings were provided with half-width barriers which drop to the accompaniment of a warning bell. The idea is that whilst they do not close the road off, the presence of the barrier, flashing lighting and the warning bell signifies the approach of a train to road users. There was normally a yellow hatched area between the barriers which further signalled to motorists and pedestrians they were not to stop in this area.

Automatic open crossings, locally monitored crossings were equipped with road traffic signals operated by approaching trains or the operation of a plunger or similar equipment, barriers or gates were not provided. This type of crossing was generally fitted to secondary lines with low maximum speeds.

Traditional Steam-Era Crossing Gates seen here at Denby Station Crossing in 1985. (*Dave Peachey*)

This photograph taken at Welbeck Colliery Junction in 1982, showcases the 'classic' semaphore signalling system. The home signal on the up line is showing a 'danger' aspect, with the light engine slowing to a stop, whilst the home and distant signals on the down line are showing 'clear' until the passing freight has cleared the section. Also note the signals for the branch, showing 'danger' as a passing MGR train can be seen on the branch to the right. (*Dave Peachey*)

'on' or at 'danger (i.e., the signal arm is in the horizontal position) and distant signals which give an advance warning to the train driver of the state of the next stop signal.

Stop signals are rectangular and are painted red with a white stripe on the front face and white with a black stripe on the reverse. Distant signals are notched at the 'free' end and painted yellow with a black chevron on the front face, and white with a black chevron on the reverse. At night, the side facing oncoming trains is illuminated via the red lens when the signal is at danger. When the signal is showing 'clear' the signal is illuminated via the green lens.

A stop signal may not be passed at 'danger' except in cases of clear failure or when specially authorised.

Colour Light Signalling

From the 1950s, colour light signalling started to be widely introduced, displacing semaphore signalling on most major routes by the 1980s. However, Semaphore signalling could still be widely found into the 1990s, and even today it is still in use on minor routes and of course heritage railways.

The first kind of system that would be recognised today as colour light was installed

Double track three and four-aspect colour light signals. (*Jamerail/Flickr*)

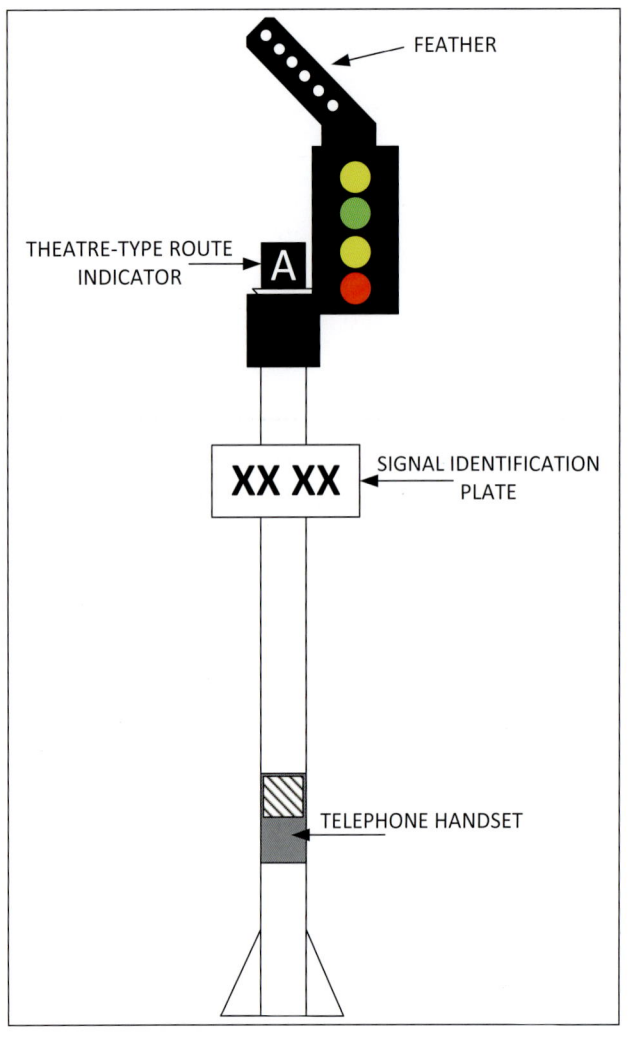

Colour light signal basic nomenclature.

onto the Liverpool Overhead Railway in 1920, the Metropolitan Railway in 1921 and the Great Central Railways Wembley Loop in 1923. All three of these systems shared a common factor in that made colour light signalling particularly useful, intensive service with close gaps between trains.

The layout of colour light signalling ranged from very simple two-aspect signals showing a red for danger and a green for clear, through to a three-aspect system whereby a yellow 'caution' light was additionally used to warn the driver that the next signal was at danger. This evolved further to the four-aspect system used by the Southern Railway in 1926. Here, the yellow caution aspect was joined by an additional yellow aspect, which gave the driver advance warning that the next signal would show a single yellow, and the signal after that would of course show a danger aspect.

Unlike the semaphore system, there are no 'distant' signals *per se*, each signal is the next signal, i.e., 'distant' by definition. As with most things on the railway, there are exceptions; in some areas the 'distant' semaphore would be replaced by colour lights with two aspects, yellow (caution) and green (clear) above it. These were installed in the rear of stop signals as so-called 'break section' signals, whereby a signal box and its associated semaphores would

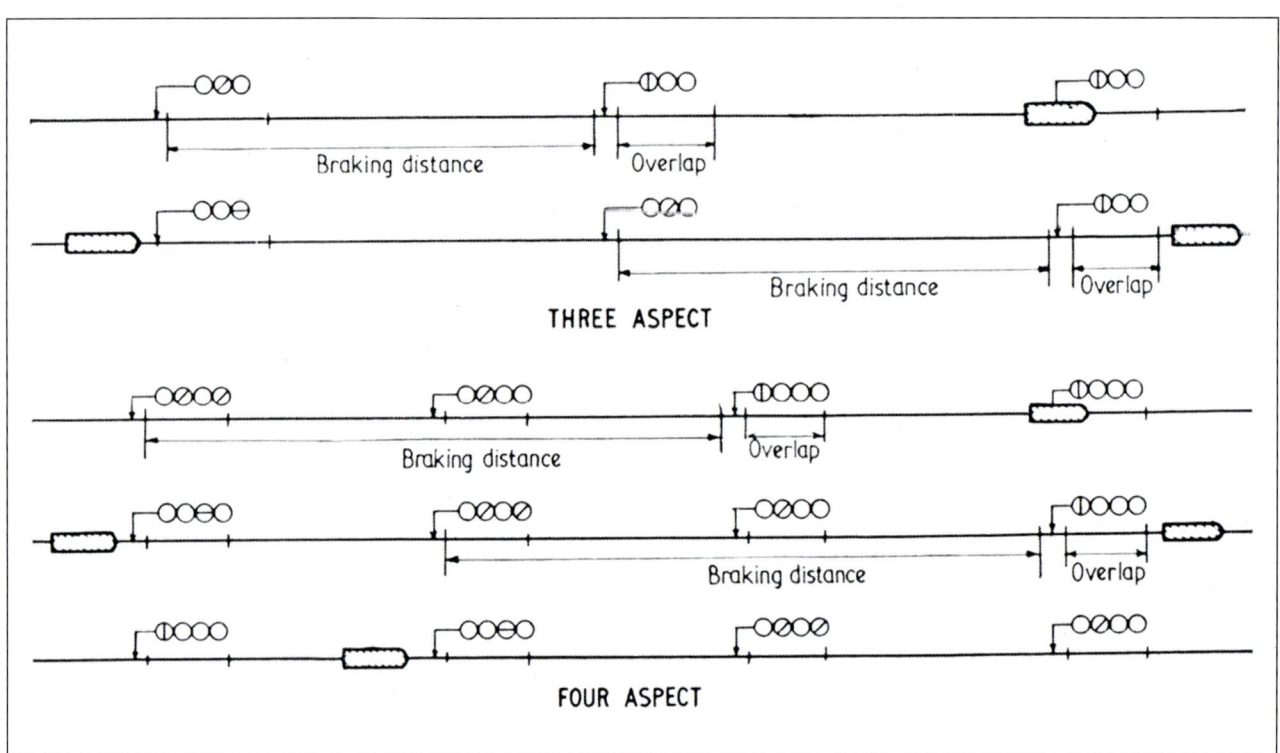

Spacing and aspects shown by three and four-aspect colour light signalling. (*British Rail*)

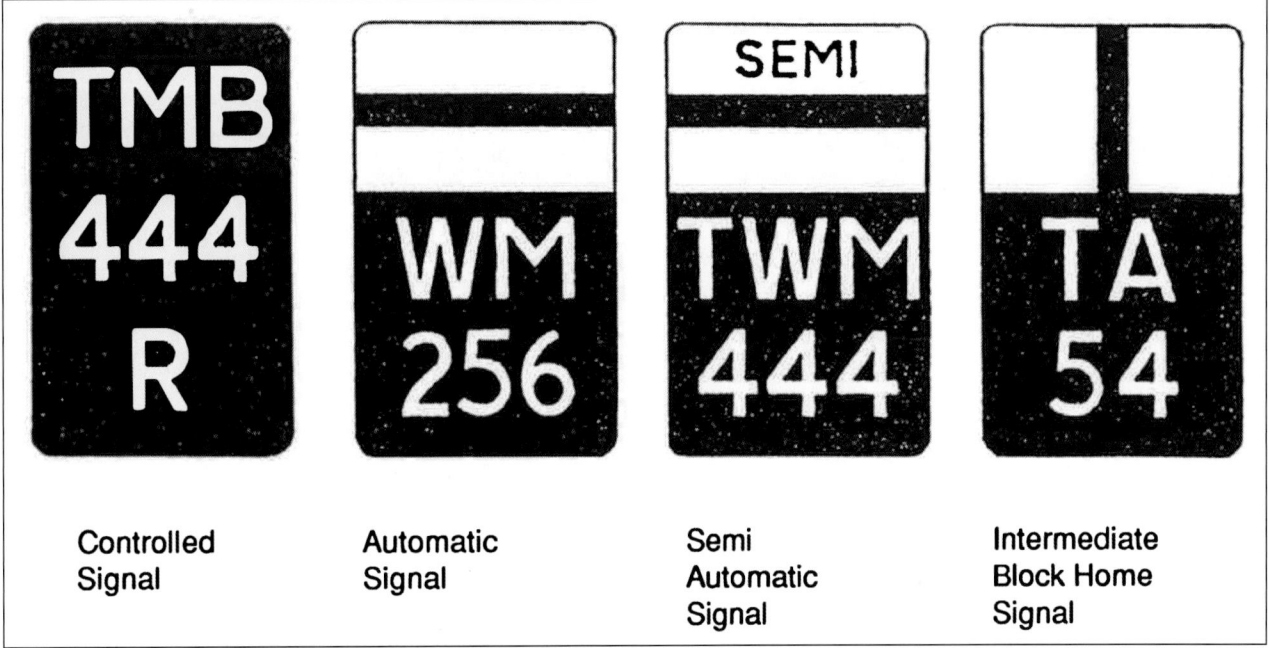

Colour signal signs. (*British Rail*)

be directly replaced by colour light signals doing exactly the same job only automatically.

Originally, colour light signals were intended to be read in the same way as their semaphore counterparts by traincrew, so at junctions a signal would be provided for each arm of a diverging route. However, after a spate of mishaps, it was quickly realised that in order to reduce the chances of a driver passing a signal at danger, a new type of junction signal would be required. The resignalling of Waterloo in 1935 resulted in the introduction of a 'feather' consisting of a row of white lights arranged at an angle and attached to the main signal, which would indicate the diverging route to which the signal applied. Note that the feather would not be illuminated if the signal was at danger, to prevent the driver seeing the route indicator and passing the signal at danger.

The resignalling of Waterloo also introduced a concept known as 'theatre type' indicators which were arranged in a square formation and were used to show numbers or letters corresponding to specific routes or platform numbers.

By the mid-1970s, the introduction of the High-Speed Train meant that drivers needed advance warning of a junction set for a diverging route to prevent the train from reaching the junction at a speed too high for the set route. The solution was the introduction of flashing yellow aspects which showed a double, then a single flashing yellow aspect before the junction signal (which would normally be showing caution) would become visible to the driver.

Reading and Understanding Prototype Signalling Layouts

An obvious source of information, particularly if you intend to construct a layout based on a real location, is a signalling layout that was drawn by the real railway for the time period you are modelling.

However, to the uninitiated these layout drawings can be a confusing jumble of symbols and meaningless information. Therefore, if we are to make a model of the real thing using a real signalling layout, then it is necessary to understand the symbols used on layout drawings.

The symbols used for both semaphore and colour light signals are detailed in BS 376 which was first issued in 1951. The standard is still current and the symbols detailed are used on most signalling layout diagrams.

The standard has tables giving symbols for colour light, semaphore and shunting signals and other signalling equipment. For a full detailed list of each symbol, it will be necessary to refer to BS 376. However, for modelling purposes, the most common symbols shown in the diagram below should suffice for all but the most complex signalling layouts.

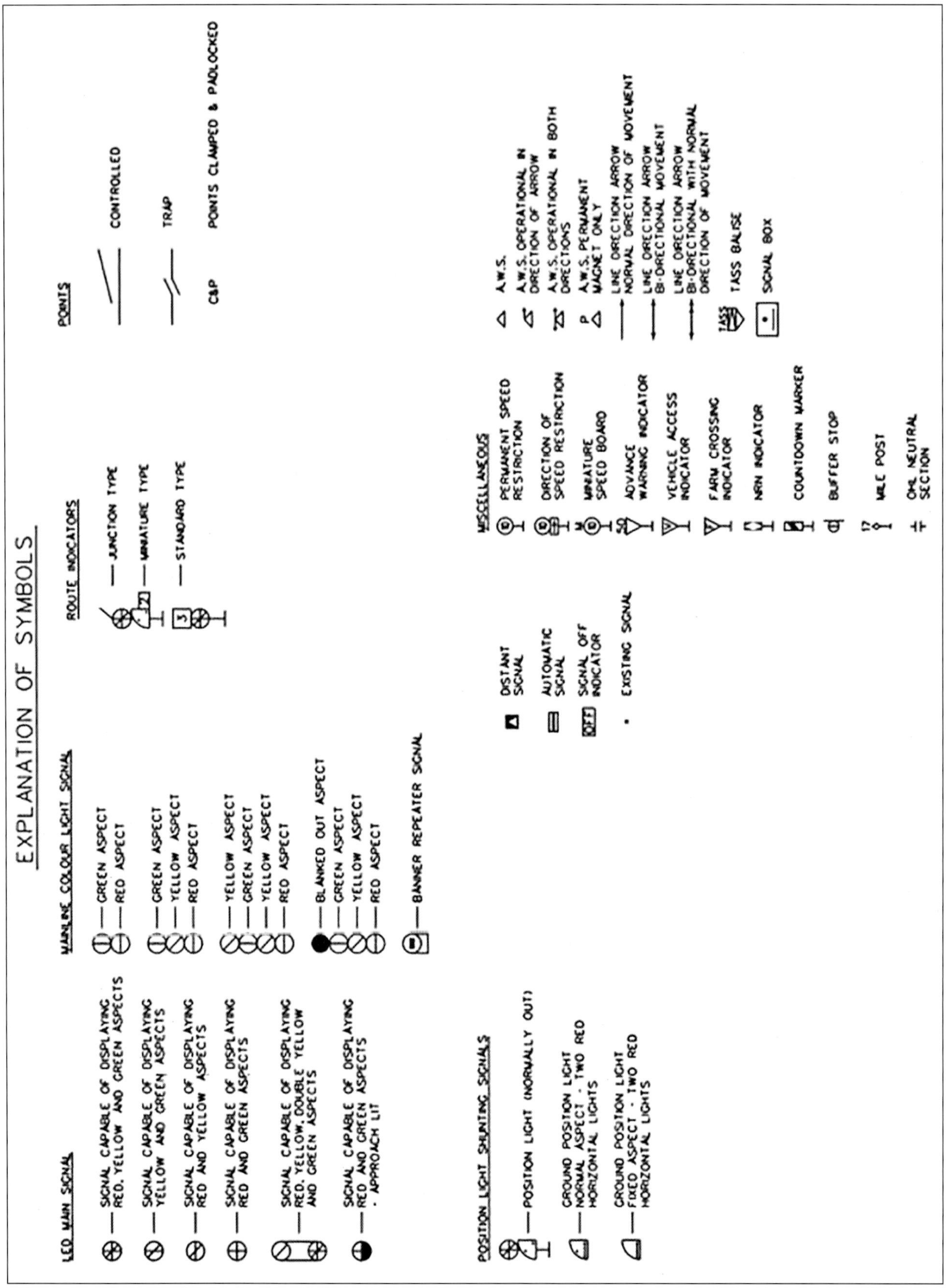

EXPLANATION OF SYMBOLS
LED MAIN SIGNAL
SIGNAL CAPABLE OF DISPLAYING RED, YELLOW AND GREEN ASPECTS
SIGNAL CAPABLE OF DISPLAYING YELLOW AND GREEN ASPECTS
SIGNAL CAPABLE OF DISPLAYING RED AND YELLOW ASPECTS
SIGNAL CAPABLE OF DISPLAYING RED AND GREEN ASPECTS
SIGNAL CAPABLE OF DISPLAYING RED, YELLOW, DOUBLE YELLOW AND GREEN ASPECTS
SIGNAL CAPABLE OF DISPLAYING RED AND GREEN ASPECTS · APPROACH LIT
MAINLINE COLOUR LIGHT SIGNAL
GREEN ASPECT
RED ASPECT
GREEN ASPECT
YELLOW ASPECT
RED ASPECT
YELLOW ASPECT
GREEN ASPECT
YELLOW ASPECT
RED ASPECT
BLANKED OUT ASPECT
GREEN ASPECT
YELLOW ASPECT
RED ASPECT
BANNER REPEATER SIGNAL
ROUTE INDICATORS
JUNCTION TYPE
MINIATURE TYPE
STANDARD TYPE
POINTS
CONTROLLED
TRAP
C&P
POINTS CLAMPED & PADLOCKED
POSITION LIGHT SHUNTING SIGNALS
POSITION LIGHT (NORMALLY OUT)
GROUND POSITION LIGHT NORMAL ASPECT - TWO RED HORIZONTAL LIGHTS
GROUND POSITION LIGHT FIXED ASPECT - TWO RED HORIZONTAL LIGHTS
DISTANT SIGNAL
AUTOMATIC SIGNAL
OFF
SIGNAL OFF INDICATOR
EXISTING SIGNAL
MISCELLANEOUS
PERMANENT SPEED RESTRICTION
DIRECTION OF SPEED RESTRICTION
MINIATURE SPEED BOARD
ADVANCE WARNING INDICATOR
VEHICLE ACCESS INDICATOR
FARM CROSSING INDICATOR
NRN INDICATOR
COUNTDOWN MARKER
BUFFER STOP
MILE POST
OHL NEUTRAL SECTION
A.W.S.
A.W.S. OPERATIONAL IN DIRECTION OF ARROW
A.W.S. OPERATIONAL IN BOTH DIRECTIONS
A.W.S. PERMANENT MAGNET ONLY
LINE DIRECTION ARROW NORMAL DIRECTION OF MOVEMENT
LINE DIRECTION ARROW BI-DIRECTIONAL MOVEMENT
LINE DIRECTION ARROW BI-DIRECTIONAL WITH NORMAL DIRECTION OF MOVEMENT
TASS
TASS BALISE
SIGNAL BOX

Common signalling symbols. (*British Rail*)

Siting and Control of N Gauge Signals

What We Need to Model

If we are to accurately represent a slice of the real railway, then the signalling system needs to be properly represented unless, of course, you are modelling a rare location where signalling wasn't in use.

To determine how to lay out a signalling system for your layout, a degree of research may be required (made easier if you are modelling a prototype location), but if the general principles outlined in this section are followed then you wouldn't be too far out.

Unless you are fortunate enough to possess a large amount of space to enable you to construct an exact scale replica of a prototype location, then it is generally accepted that a degree of adaptation and compression will be required. This is entirely achievable, and the result can still be aesthetically pleasing and prototypically correct, although it does require us to be somewhat creative.

To start with, choosing the correct type of signals for the period and location being modelled is vital. By the 1970s, pre-grouping signals had largely disappeared from the scene. However, it was not an unusual to see quite ancient hardware even into the 1980s. It would be entirely feasible, for example, to see a pre-BR GWR signal post with a BR signal head and perhaps a signalbox telephone attached to it. However, unless you are scratchbuilding (a whole subject in itself), it's accepted that the modeller will be restricted to those signals that can be readily obtained from the local model shop, therefore the choice may be made for you.

The actual design of a semaphore signal could depend on the railway company that originally installed it. We should be mindful of the basic differences between, say, a GWR and an LMS signal. A real blow to accuracy would be to have a layout set in the Northeast signaled with GWR type signals, for example.

By the 1980s, most principal lines were signaled with colour light signals. With a few exceptions, colour light signalling was more standardised and less region-specific than semaphores.

Signalling for Model Railways

The conditions we might find on our model railways are appreciably different to those on a real railway system. Principally, our driver is not inside the cab of a locomotive and therefore can communicate with the signaller directly (assuming of course, that the driver and signaller are not one and the same person, which they frequently are on most layouts). Furthermore, an accident or derailment on a model is rarely more than a minor nuisance, clearly this is not the case on the prototype.

Before we discuss the application of signals on a railway layout, it would be beneficial to identify situations where signals were *not* required, these being:

- Within sidings (with a few exceptions), as movements within sidings are made at low speed with groundstaff present.
- Smaller industrial or private railways with limited traffic and low speeds

Disc type shunting signals seen here at Abbey Foregate Junction in 1987. (*Jamerail/Flickr*)

Under the traditional 'analogue' control used on model railways, the electrical section used to keep trains apart from one another corresponds with the block quite closely. If we were to allow two trains into the same section, chaos and possibly damaged models would ensue as two trains would be under the control of one control unit. The advent of Digital Control has removed this limitation somewhat, it is now possible for two trains to be operated very close together more akin to the prototype, however the implementation of a block system is still beneficial to ensure two trains cannot be run into one another, with the attendant inconvenience and possible damage this may cause.

If we break our layout into block sections, it becomes easy to control trains in a prototypical fashion.

Electric and electro-hydraulic semaphore signals were not common on the railways prior to the 1980s. Mechanical signals require a signal box and if a level crossing was required then a logical move was to place the signal box there, so the signalman could operate the crossing gates as well as the local signals. It would be unusual to see a gated level crossing without a signal box next to it but in some locations, they had a small cottage for a crossing keeper. If the road crossing the railway was at all busy, such crossings were often replaced by a road bridge from the 1960s on.

Colour light signals do not require a signal box but disused signal boxes existed in many locations long after they were closed, particularly in the vicinity of stations and crossings.

This signal has several points of interest. Note the combination of semaphore and colour light signal, a combination increasingly rare by the 1980s, but, more importantly, the white contrast board behind the signal to aid driver visibility. This is an excellent example of the types of modifications sometimes needed to ensure that the signal was visible in all conditions. This particular signal was the Up Harwich Section Starter controlled by Mistley signal box near Colchester, taken in 1984. (Jamerail/Flickr)

Signal Sighting

Given the safety critical role that signalling plays in the operation of the prototype railway system, it was obviously essential that signals were positioned in a manner that allowed drivers to view them clearly. 'Sighting' as it is known involved not just the position of the signal, but it could also influence the design of the signal chosen for that particular location, for example repeater arms, which reinforced a particular signal, could be used in areas where the main signal might not be seen.

Curves, overbridges and tunnels could pose significant challenges for engineers when deciding where to position signals, and in some cases, special signal designs were developed to overcome a specific challenge. A very common special signal on semaphore layouts was a tall signal post carrying two co-acting signal arms; one arm was placed high up and the other was placed at eye level. These were often found where an overbridge was to be found on the approach to a signal. A signal was thus developed that could be seen both above and below the bridge.

The essential point is that the modeller should essentially 'sit' inside their model locomotive when siting signals and attempt to visualise what the fictional crew of their miniature locomotive would see from the windows of the driving cab. Essentially, this is the same principle followed by engineers on the prototype, so a modeller placing their signals where they would be most readily seen by traincrew will come the closest to replicating prototype practice. This remains true even if established rules such as having signals out of position – for example a signal placed on the far as opposed to the near side of the track – are broken. A review of photographs taken of the real railway will evidence a surprising lack of standardisation, particularly as the requirement for signals to be easily seen trumped most other procedural matters as this was considered a principal matter of safety.

Siting and Modelling of Ancillary Signalling Equipment

There are a number of items of equipment associated with the signalling system which are prevalent in a typical railway scene. These are visually obvious; point rodding for semaphore signalling systems, trackside electrical cabinets, concrete trunking and under track protective tubes for signal cabling and so on. There is little to be gained from a painstaking replication of a prototypical set of signals if we are to ignore these important secondary items of equipment.

The exact layout and configuration of ancillary signalling equipment would be the subject of a number of technical drawings and, in many cases, would be unique to the location. As ever, careful review of period photographs, either of the location in question, or, if modelling a fictitious layout, a comparable location, will yield the best results. The goal should be an impression of the real-world installation, not necessarily a painstaking replica.

This photograph, taken at Desford Colliery Sidings in 1983, shows the typical rodding associated with a facing point lock. Note the wire for the semaphore signalling on the right of the picture. (*Jamerail/Flickr*)

Duncan Hunnisett's 'Shirebrook' demonstrates that an impression of signalling equipment, rather than a painstaking replica can still be visually stunning. In addition to the signals and the box, a representation of point rodding, trunking and electrical cabinets complete the scene. (*Duncan Hunnisett*)

Train Protection Systems

Train protection systems introduced to the UK network are, sadly, often the result of investigations into accidents which have cost many lives. The primary purpose of train protection systems is to apply the brakes if a train passes a signal that is showing a danger aspect. During the BR Blue and Sectorisation eras, the two systems in use were the AWS in use from 1950 onwards and the Automatic Train Protection (ATP) system which was in limited use from 1988 onwards.

Automatic Warning System

AWS are not new, systems in varying forms having been proposed, theorised, patented and introduced as far back as the 1850s. However, for the purposes of this book, we are concerned with the AWS system introduced by British Railways in 1950. The introduction of AWS was given renewed urgency after the 1956 Harrow and Wealdstone crash.

AWS initially provided train driver with an audible warning and a visual indication that they were approaching a distant signal that was showing a caution (yellow) aspect. The scope of AWS was later extended to give additional warnings for:

a) A colour light signal displaying a double yellow (steady or flashing), single yellow or red aspect;
b) A reduction in permissible speed;
c) A temporary or emergency speed restriction;
d) An automatic barrier crossing locally monitored, an automatic open crossing locally monitored, or an open crossing.

AWS is designed to ensure with absolute certainty that the driver of a train receives a warning if they fail to see a distant signal, sufficient to take any action should the signal have a stop action against them. It achieves this by first providing an audible indication of the location and setting of each signal as it is approached. Secondly, a visual reminder that a caution aspect has been seen and acted upon, which stands until the next distant signal is reached. However, AWS is not a train protection system, it merely provides an audible and visual warning on approach to a signal of the aspect that is displayed at the precise moment that the train passes over the trackside equipment.

AWS uses inductors placed between the running rails. The inductors consist of a permanent magnet followed by an electromagnet (in the direction of travel) which is energised if the linked signal is showing a proceed aspect (green). If the signal is showing a caution or red aspect, the vehicle detects that only the permanent magnet is energised and sounds warning in the driving cab. If this is not cancelled by the driver within three seconds, the brakes are automatically applied.

To protect the equipment from any loose part of a train, an obstacle deflector was fitted to the back of the installation (or both ends if the line was bi-directional). The obstacle deflectors, with their distinctive profile, gave rise to the colloquial term 'ramp' which is used by many enthusiasts.

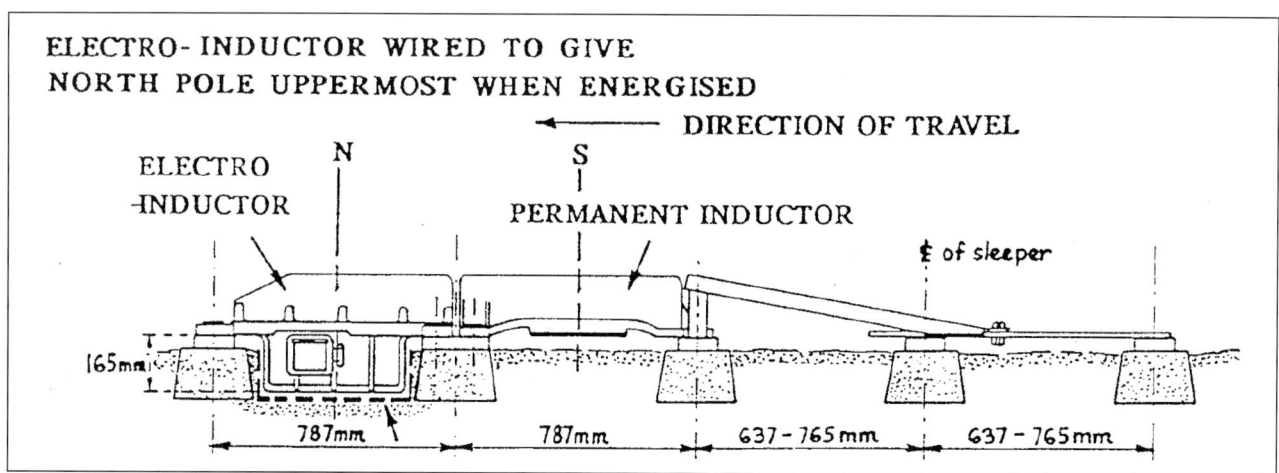

AWS inductor typical construction and nomenclature. (*British Rail*)

Originally, the AWS indicator inside the vehicle was a 'sunflower' visual indicator and a separate bell. However, this was changed on newer vehicles to a combined alarm/indicator unit with a horn. The origins of the 'sunflower' visual indicator are obscure, but one theory has it that the sunflower was only meant to be used for testing, but having appeared in numerous photographs, it was kept for publicity purposes.

Clearly, in N gauge, equipment inside vehicles would be impossible to represent. However, trackside AWS apparatus, particularly the distinctive yellow inductors are often conspicuously absent, especially in N gauge. This is easily remedied, as fairly accurate representations of AWS inductors are available. These can be fixed to most proprietary and hand-built track. There are also cottage industry suppliers producing AWS inductors as 3D prints.

BR documentation stated that any semaphore distant signal and any colour light signal capable of displaying a single or double yellow aspect on a line that met the criteria for AWS was to have a track inductor installed. In addition, inductors were to be fitted at reflectorised distant boards, AOLC boards and permanent and temporary speed restriction boards.

The standard distance between the signal and the end of the ramp was 180m on the approach side of the signal. This translates to approximately 1.2m in N gauge. This distance can be reduced to fit within a compressed layout space whilst still looking visually realistic. Additionally, placement on the real thing varied considerably, and inductors were sometimes closer to the signal than stipulated in official documentation.

Finally, official BR documentation stated that inductors should not be closer together than 270m to avoid confusing drivers. Therefore, in N gauge, you should avoid grouping AWS inductors within 185cm of one another on the same line. This only really becomes an issue on large layouts, as most layouts will only feature a handful of inductors.

Automatic Train Protection

A number of signal passed at danger incidents (SPADs) in the 1980s caused concern amongst British Rail management and led to calls for a system that would prevent SPADs entirely. After the Clapham Junction accident in 1988, British Rail was tasked with introducing a new train protection system within five years. It was recognised right from the start that such an undertaking would be difficult and expensive.

In 1988, BR started a three-year development project, with the goal of starting implementation in 1992. The new system was designated ATP and was considerably more advanced than the AWS system. The principle aim of the ATP system was to take control of the train if a signal was ignored or missed by the driver.

ATP used an onboard computer which was pre-programmed with the vehicle's operational parameters. This computer received data regarding the status of the route ahead from trackside equipment. Data passed to the train by such equipment included maximum permitted line speed, signal aspects and distance and gradient information. This data was used to supervise the safe operation of the train. The system did not automate the driving of the train, but if, for example, the system detected that the train was operating above the maximum permitted speed, it would intervene and reduce it.

The first two routes selected to trial ATP were the Great Western and Great Central lines out of London Paddington and London Marylebone respectively. These two routes were fitted with different trackside equipment (principally the ATP beacons placed in the centre of the track were visually different), but in terms of operating principle were the same. Owing to the huge costs involved, and the privatisation of the network in the mid-1990s, these routes were the only two to ever be equipped with ATP, with the train protection and warning system being implemented by Railtrack early in this century.

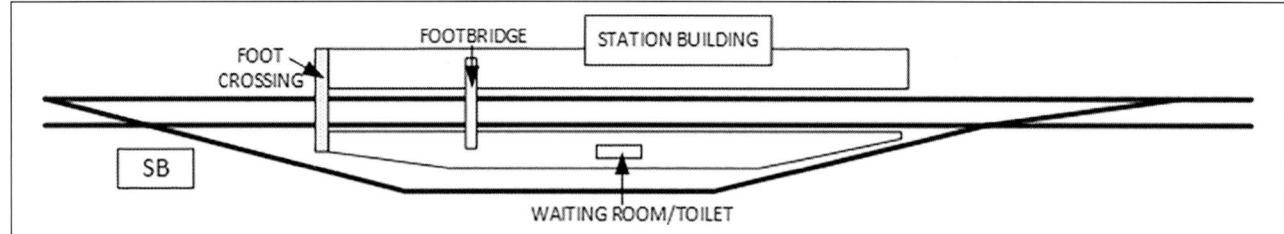

Loop road served by island platform.

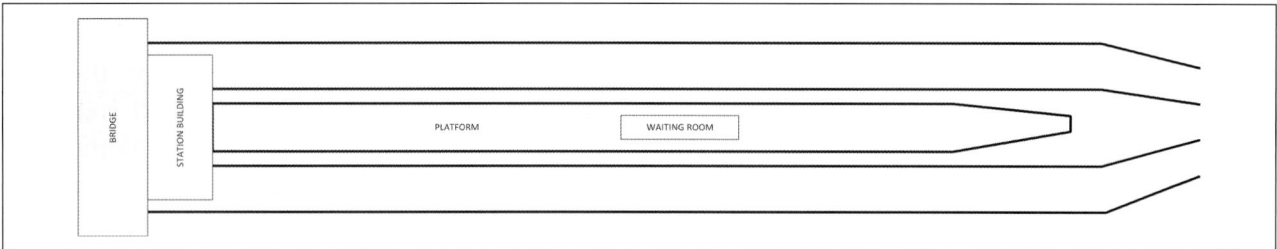

Island platform with road overbridge facilities.

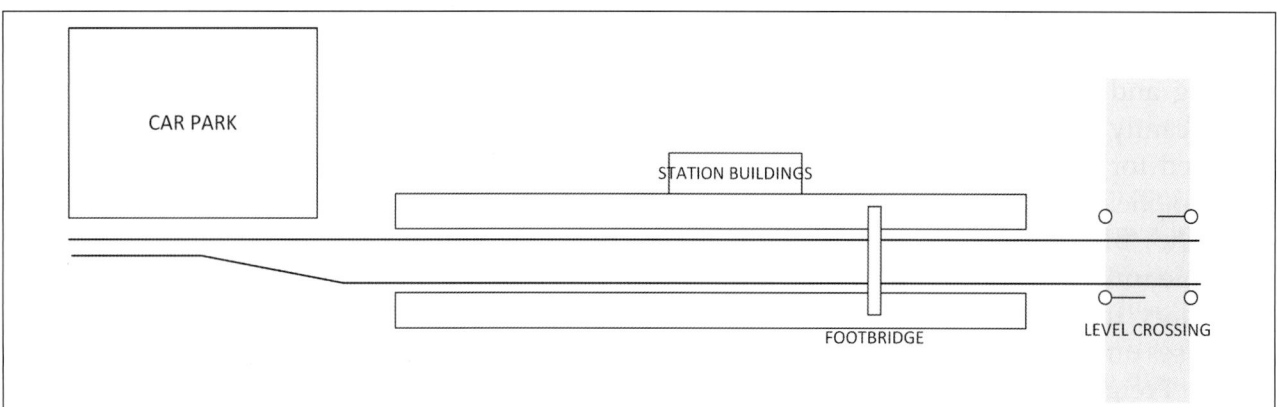

Small to medium size suburban through station.

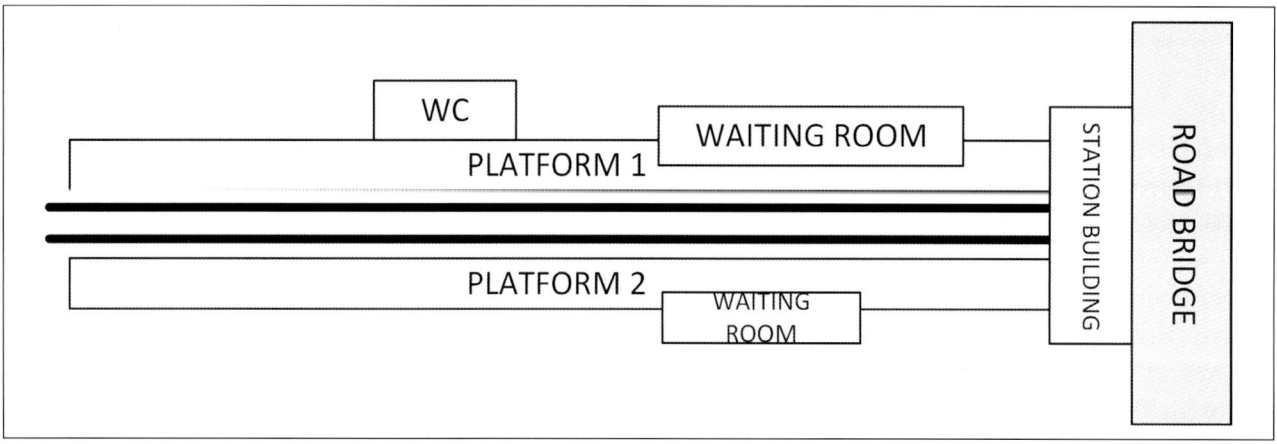

Small through station.

It was realised in the Victorian era that trains waiting at stations were holding up other trains and causing a bottleneck on busy routes. One of the solutions adopted was placing loops either side of the running lines in which trains could wait. Locomotives could also use the loop to run around the train and set off in the opposite direction. Such station layouts were common even into the 1990s.

The centre island with road level station building was particularly favoured by railways such as the Great Central and the Midland. The design allowed additional avoiding lines located outside the platform roads if additional capacity was required. These lines would be used by passing passenger and freight trains alike.

A typical small to medium sized suburban through station would probably see plenty of commuter trains, with the occasional non-stopping express or freight. By the late 1970s, most station layouts of this type had undergone significant modification and rationalisation. The former goods yard was often turned into a car park and any goods facilities demolished. The level crossing might have survived, although it would have probably been converted to

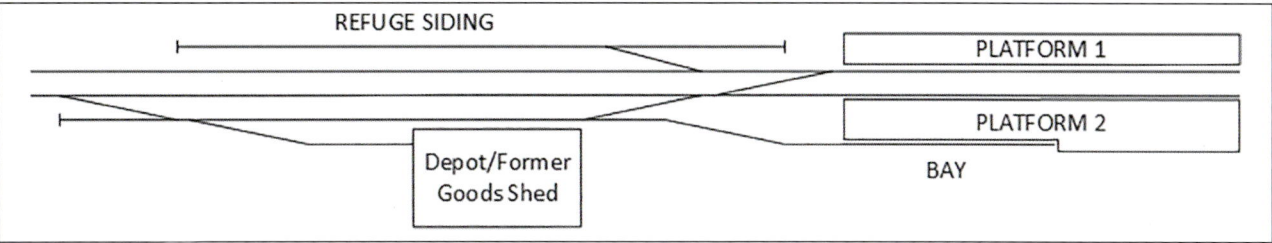

Through station with sidings.

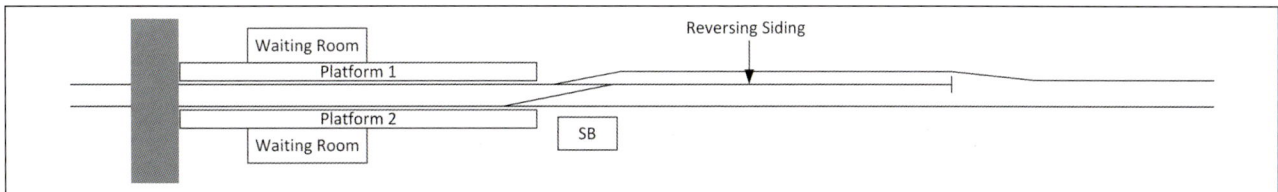

Through station with reversing siding for multiple unit operation.

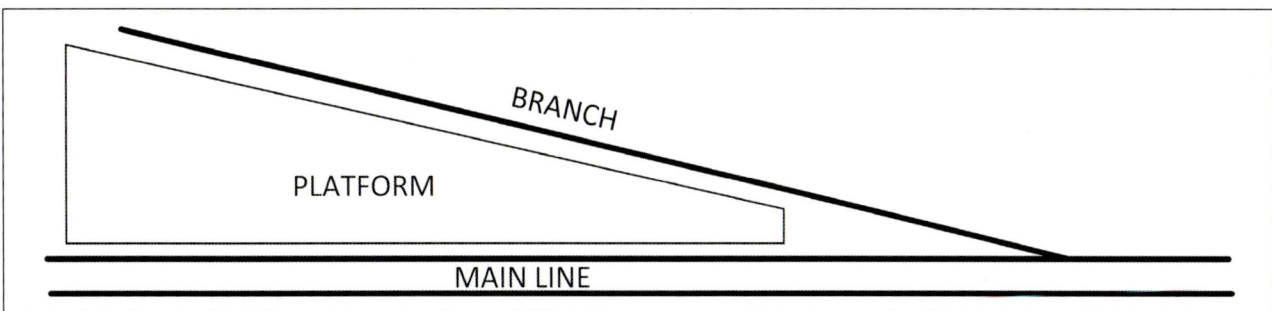

Island station with platform face for branch.

Dovey Junction Station is situated on the Cambrian line between Machynlleth and Borth. It is a classic island platform Junction station, with the line to Penhelig branching off to the left. (*David Flitcroft*)

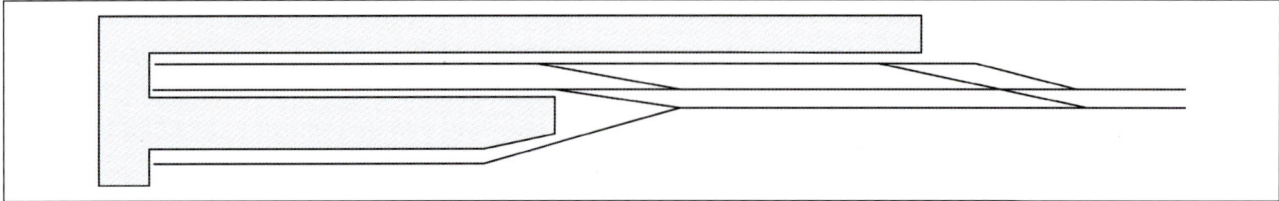

Terminus with two train capacity roads.

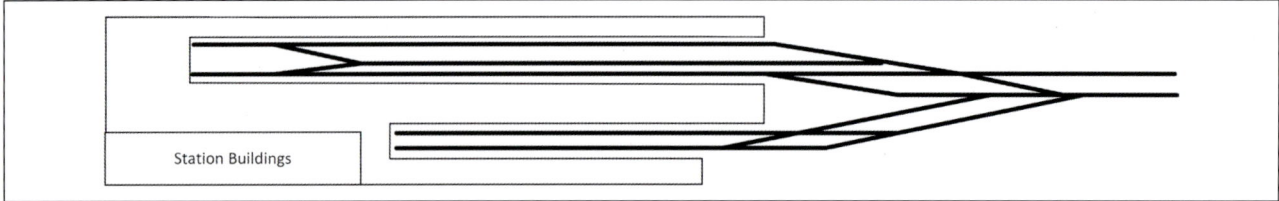

Terminus with variable length platforms.

Perhaps the classic Scottish Terminus, the station at the Kyle of Lochalsh, photographed here in 1973, would make an attractive model. (*John Ford*)

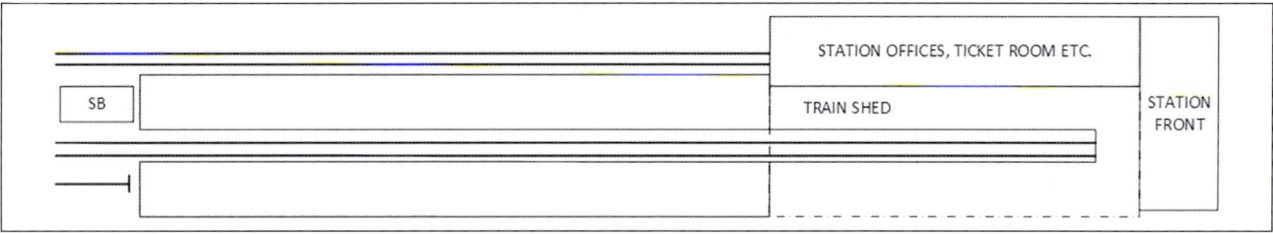

Features required to give an impression of a main line terminus.

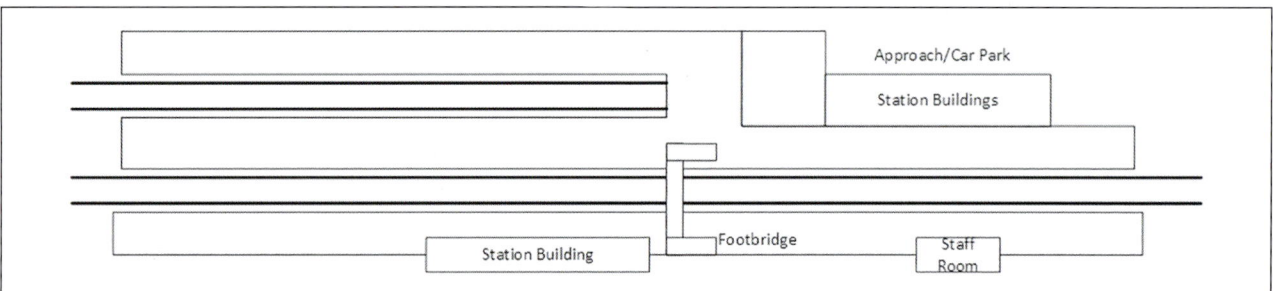

Through terminus.

automatic operation. In rural areas or on lightly used lines, Semaphore signalling and attendant signal box may have survived, but it's equally likely the signal box had been demolished (or perhaps left disused) and the signals upgraded to colour light signals.

Even as late as the 1990s there were still some stations with sidings, long after the goods yard had closed. The inclusion of a refuge or reversing siding (for operation of multiple units) can add operational interest to an existing through station. In addition, some locations had the former goods facilities repurposed into other rail-served industries, such as a parcels depot.

Many stations were situated at junctions. These were often island type platforms, with one platform face serving the 'main' line and the other serving the 'branch'. Some stations also had an additional platform to serve mainline passengers boarding and alighting trains in the opposite direction.

The terminus is a station type that makes a regular appearance on layouts, probably because it only requires a scenic break at one end of the layout. A terminus could vary from a tiny station at the end of a branch of the type favoured by GWR modellers, to a large multi-platform city terminus such as Waterloo or St Pancras. It is possible, with some careful thought, to give the impression of a large terminus on a layout without modelling the station in its entirety.

Another, more unusual, layout was the through terminus, which incorporated features of both a through-station and a terminus, with bays adjacent to the through platform. This was common where a branch service was required to terminate in and otherwise intermediate station.

Loops and Refuges

An important manoeuvre on a railway, both real and model, is the passing of trains travelling in opposite directions on a single line. To achieve, this, a length of track known as a passing loop is required. Often, but by no means universally, such loops would be provided at a station, with a facing platform on each side of the loops for the use of passengers, the function of which should be obvious. A passing loop allows one train to pull into a station and embark or discharge passengers, whilst the other train can either come into the station (on the opposite platform) or pass straight through without further encumbrance.

In the UK, the traditional method by which slow moving freight trains could be moved out the way of priority passenger trains was via the use of refuges and relief sidings. This was a long siding in the trailing direction, into which a slow-moving freight train could be backed to wait for the faster passenger train to pass. Moving into such a siding was a slow laborious process and was not particularly popular with crews. A refuge or relief siding is rarely modelled on a layout because the distance between two terminal points on even a large layout is insufficient to warrant the inclusion of such a feature. However, an intermediate signal box with attendant refuges as shown in the diagram below could be an interesting feature on a larger continuous run type layout, offering the potential for increased operational interest.

Another type of loop known as a lie-by loop was found on busy double track main lines. The lie-by loop permits a goods train to be quickly moved out of the way of a passenger train.

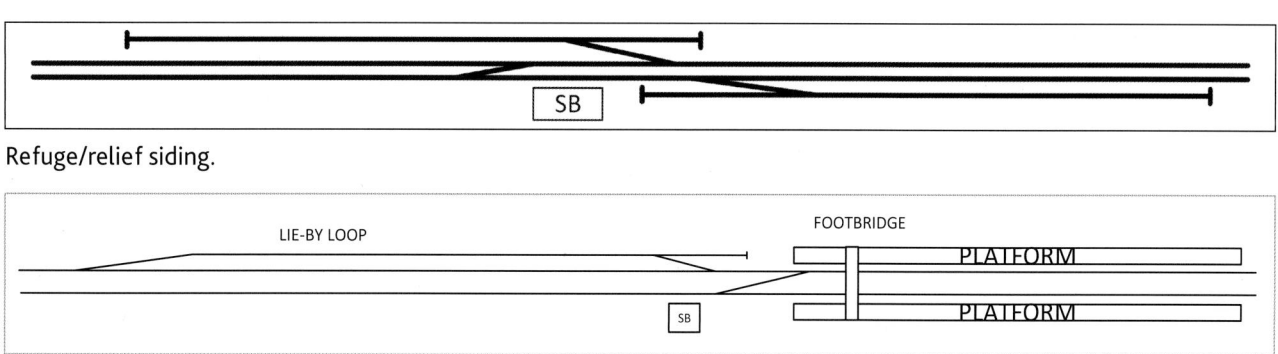

Refuge/relief siding.

Station with lie-by loop.

Sidings and Yards

Ladder sidings are simple to construct and visualise and are popular with modellers. However, the disadvantage is that anything more than four or five roads leaves the top siding increasingly short, which can cause operational difficulties. A more sensible arrangement is known as a compound lead, and, whilst more complex, allows for sidings of roughly equal length, whilst only taking up a little more space than a ladder siding.

Hump shunting yards are visually impressive, both in model and prototype form. However, they are not particularly interesting to model, and in a small scale such as 'N' it is difficult to make the wagons 'roll' down the siding convincingly due to weight. Such a yard also has the downside of needing a large amount of space in order to look convincing.

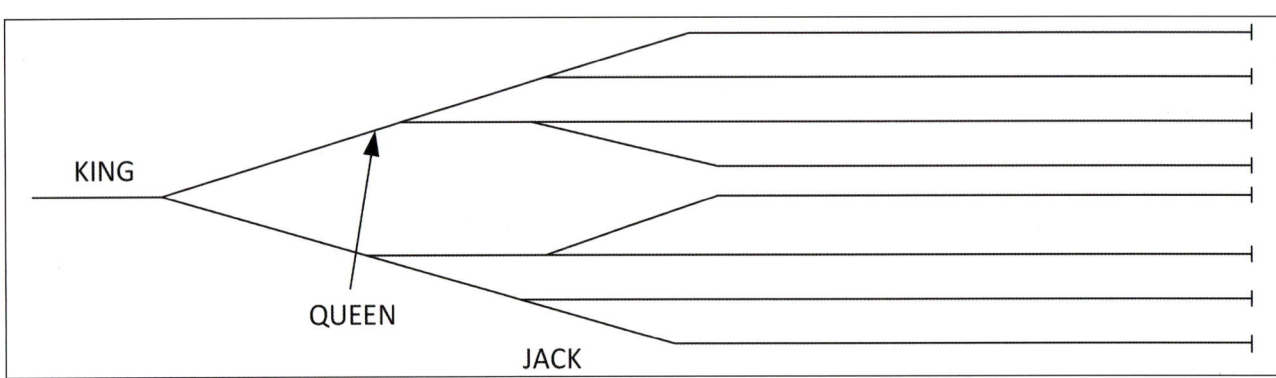

Compound lead for sidings.

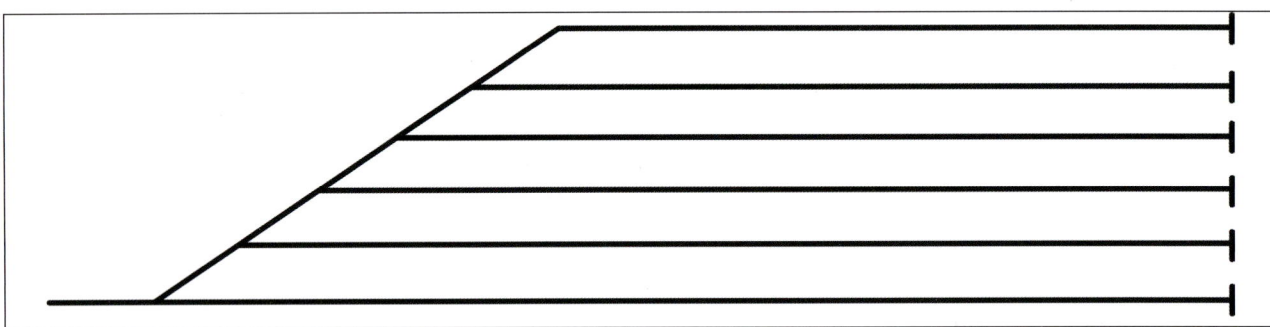

Ladder sidings.

The yard at Tinsley was a famous example of a hump shunting yard. This photo was taken in 1986 when the yard was still busy with coal traffic. (*Jamerail/Flickr*)

For intensive shunting and marshalling purposes, a concentration yard would be more suitable. A loop serves as arrival and departure lines, there are shunting spurs and a set of parallel sidings where wagons are formed into sections or block trains for onward dispatch. You could also include a wagon repair depot, and a small traction maintenance depot (TMD) which were frequently co-located at such sites, and would add additional interest.

A container terminal can make an interesting model, although to simulate the loading and unloading of container it would be required to motorise the sited crane and provide a method of loading and unloading containers (via a magnet). The large overhead crane would lift containers from the train to waiting lorries and vice-versa. There may be two or more roads, to which intermodal trains would be shunted, the locomotive would often leave the train in-situ whilst it was loaded, another locomotive arriving later to pick the train up. A shunting locomotive would not normally be provided, use would be made of arriving main line engines to work any required moves within the yard.

If additional operating interest is required, the terminal can be sited adjacent to a main line (possibly separated by a wire fence), allowing passing freight and passenger trains, similar to the Freightliner yard that was formerly at Beeston near Nottingham.

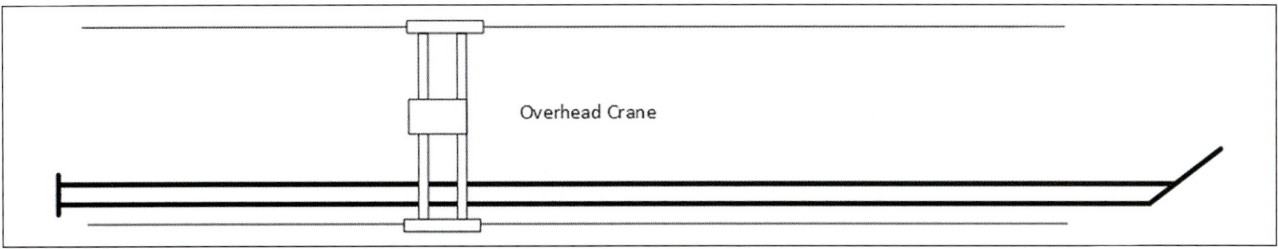

Small container yard.

A large and distinctive container crane is required for a model of a container terminal. This example, photographed at Cardiff Pengham in 1989, is fairly typical. (*Jamerail/Flickr*)

Mallaig Junction oil terminal. (*David Ford*)

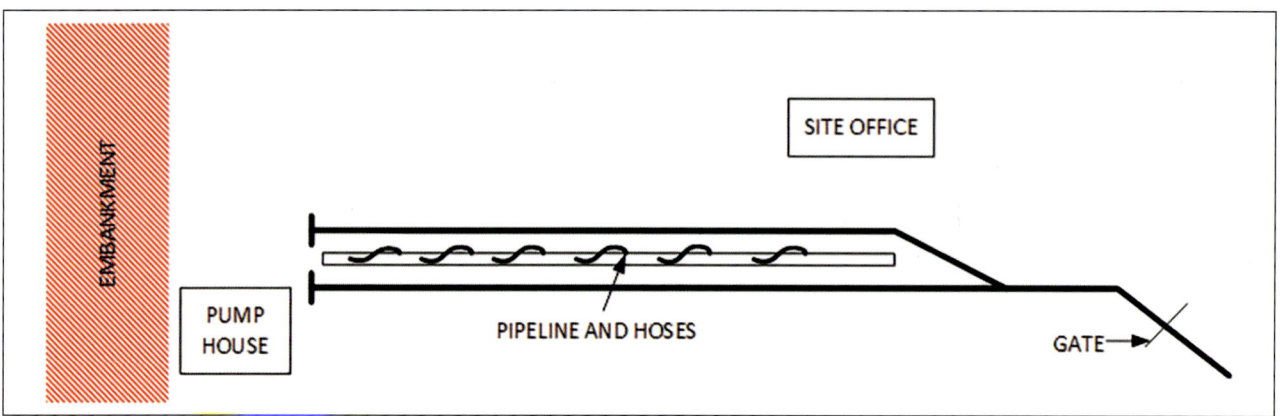

Small oil terminal layout.

Small oil terminals such as Mallaig often had gates on their boundary. This marked the point at which British Rail Regulations started/ended. Sidings such as these were often also fitted with catch points for additional protection should wagons runaway. Mainline locomotives would enter the sidings to pick up laden trains or drop off empties. In small sidings such as these, a shunter wouldn't be needed. Sometimes, in hazardous locations such as refineries, the locomotive would not be allowed to enter the yard, so a 'reach' wagon (often a cut down chassis from a van or similar) would be used to provide a gap between the locomotive and the train. The reach wagon (if present) would normally be stored somewhere near the access to the main line, possibly in a small siding or just left in the yard entrance.

Oil terminals were normally surrounded by high fences for security, and area inside the terminal was generally (although not religiously) kept free of vegetation and weeds to prevent fires. A visible feature of these terminals was the raised pipes with control valves in between the tracks, and there would normally be a control building and a store for additional hoses etc.

Exchange sidings are sidings where wagons from a private or industrial sidings are picked up and dropped off by mainline locomotives. Often, there would be a small industrial or ex-BR shunter in attendance, either tripping the wagons to and from the loading point or marshalling individual wagons into rakes for pickup by mainline locomotives.

Several large facilities for the loading of aggregate still existed in this time period. The

The exchange sidings for the Humber Refinery are shown to good affect here in this 1987 photograph. Note the small industrial shunting engines which are a prominent feature of industrial railways and should be present on all but the smallest of exchange sidings. (*Dave Peachey*)

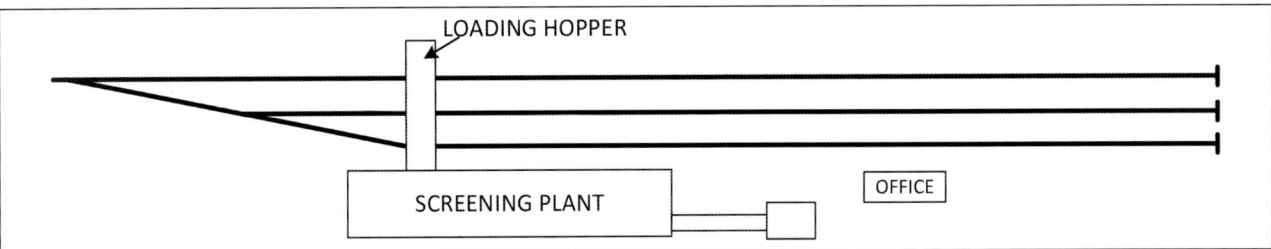
Aggregate loading plant.

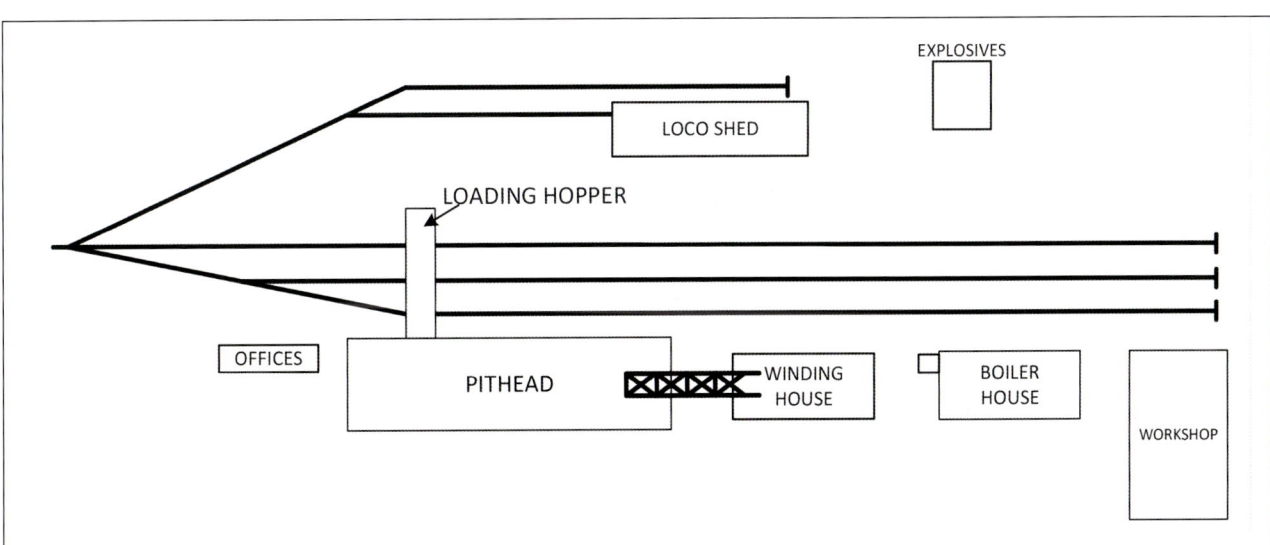
Simplified coal mine loading site with essential features modelled.

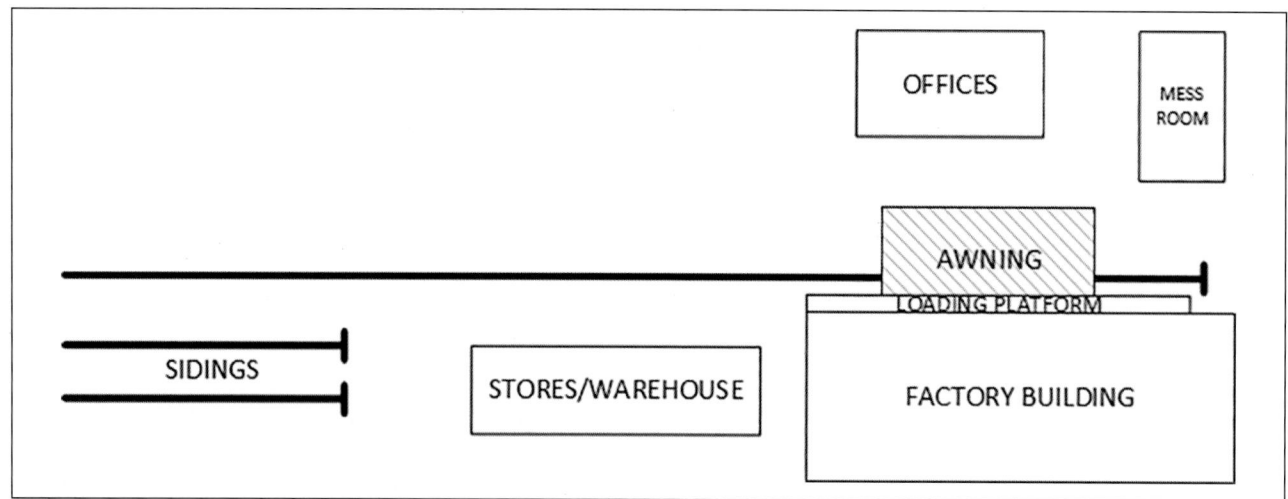

Whilst not as common as they once were, there were still examples of rail connected factories to be found into the 1980s. They might only have one or two sidings, or they might be larger, such as the works of the Butterley Engineering Company on the Erewash Valley line, which still had a rail connection well into the 1990s. The sketch above gives some basic ideas for a factory scene. This could easily be fitted into the corner of a larger layout if required.

major difference was that the larger locations had a loading facility which was fed with aggregate delivered from the quarry often in trucks. The major limitation to this type of facility is the length of the sidings required, block trains of aggregate, even in N, will take up significant space.

A coal mine, even a small one, would be a large model, even in N Gauge. However, it should be possible to construct a fairly accurate representation of a rail loading site, as this only requires a few buildings and a sensible amount of track.

Speedlink terminals were large, utilitarian structures that allowed the transfer of goods to and from railway wagons/road lorries. Most were built using private funds, normally provided by the company who required the rail service. Again, this is reasonably straightforward to model, only requiring a fairly basic industrial type shed and a few lengths of track.

Maintenance and Servicing Depots

In the steam era, even the smallest terminus station would have a single road engine shed for coaling and watering locomotives. The consolidation of servicing facilities to reflect the move to diesel/electric locomotives and multiple units saw large scale closure of many small servicing sites and consolidation of others into larger regional maintenance centres such as the huge shed at Toton. However, smaller depots, the type likely to find favour with space-starved modellers, did survive in limited numbers.

The type of depot most favoured by modellers (ostensibly so they can show off their extensive locomotive collection), is the locomotive depot, more properly termed a TMD in the diesel era. These depots can range from a simple small facility with one or two roads, to a huge multi-road shed such as Toton or Crewe with extensive facilities.

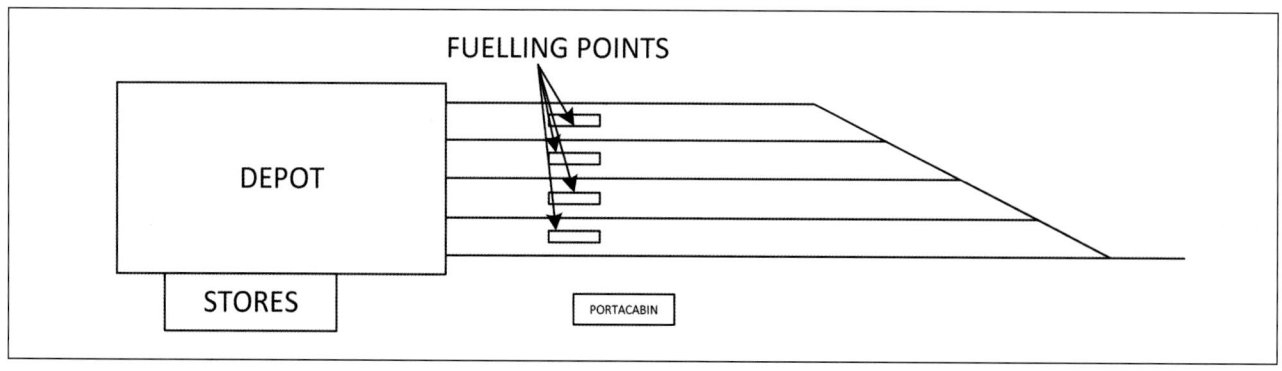

Small diesel era TMD. This layout is loosely based on that at Tinsley.

Typically, a small TMD might include a two, three or four road covered shed, possibly with raised platforms inside for locomotive servicing. There might be stores and a workshop attached to the side of the main building. There might also be an automated locomotive washer, and possibly additional temporary type cabins for accommodating infrastructure workers, or as additional office space (possibly even a mess room). There would also be a fuelling point, either one or two roads with covered fuel pumps. These would be located either next to the TMD or slightly away from the main site, and (for obvious reasons), such locations needed their own drainage and safety precautions against ignition or spillage. Some small depots, such as Tinsley, had the fuelling point immediately in front of the shed, which would make for a compact and attractive layout.

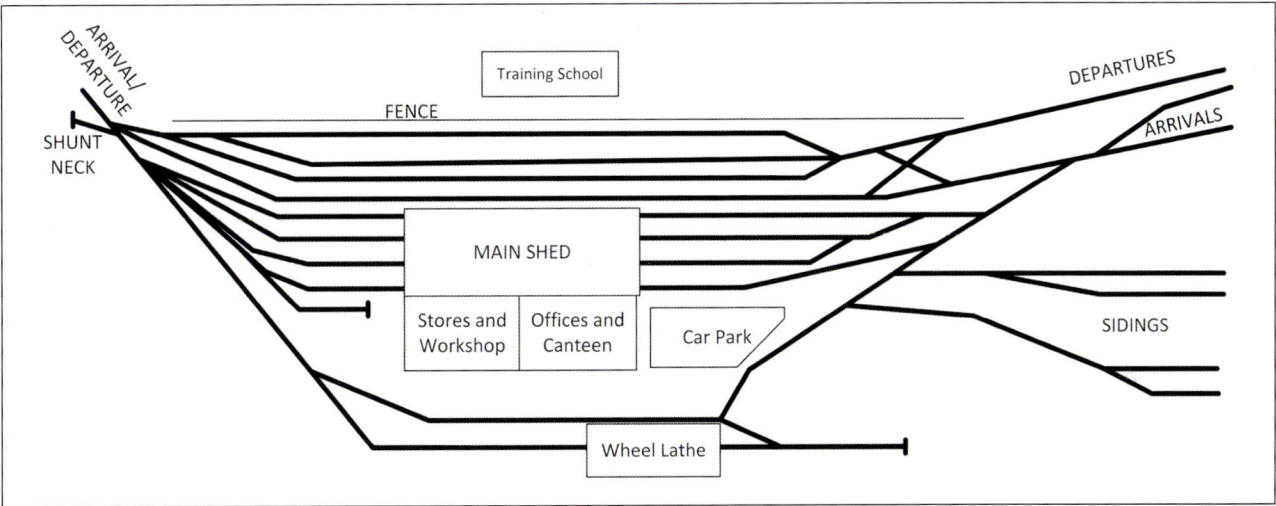

Large electric locomotive depot. This example is modelled on the Crewe Electric Maintenance Depot and would make for an attractive layout if space wasn't a consideration.

Shirebrook TMD is captured perfectly on Duncan Hunnisett's layout 'Shirebrook'. Even a small depot, well executed, can be visually stunning. (*Duncan Hunnisett*)

Chapter 7

Dressing the Scene

Trackside Clutter

The operating railway is not a sterile environment, with a great deal of material and detritus present lineside, some of it occurring naturally, some of it left by railway workers, and some dumped by ne'er-do-wells during their nefarious activities.

A look at pictures of the railway in the 1970s and 1980s will reveal a network largely in decline. Although there are many interesting facets of the railway during this period, there is no getting away from the fact the railway was under financial pressure and managed neglect was generally the norm. In addition, the withdrawal of steam locomotives in 1968

This shot taken at Hereford Goods Yard in 1987 showcases typical trackside clutter that adds to the atmosphere of the railway. Note the overgrown nature of the sidings, piles of pallets, sleepers, rails and other junk, and the areas of oil staining. (*Jamerail/Flickr*)

removed the requirement to closely manage lineside vegetation which had posed a fire risk with sparks from steam locomotives. This, coupled with underfunding, declining passenger numbers and the running down of many manufacturing and mining centres produced a cluttered, overgrown, dirty and decrepit railway system.

Adding appropriate trackside clutter to your layout can do much to enhance the atmosphere of your layout and can help tie it to the period being modelled.

Discarded Rails and Sleepers

Rail and sleepers were replaced on a frequent basis, ranging from replacement of a single rail, to large scale renewals. Owing to the low value and the size and weight of rails and sleepers, it was common for worn out components to be left lineside, ostensibly for future pickup but in practice rarely retrieved.

Cable Drums

The railway used huge quantities of cabling. It should therefore come as no surprise that cable drums (both full and empty) could often be seen lying around the lineside.

Ballast Bins

Ballast bins were built to provide the track gang with a source of material for minor repairs. Some ballast bins were constructed from brick, but the majority were constructed from wood, often in the form of old sleepers. In later years, precast concrete ballast bins became a common sight.

Ballast bins continued to be a visual part of the scene throughout the 1970s and 1980s, but by the 1990s they had been made largely redundant and the examples that remained were often in a dilapidated state and overgrown with weeds.

Signs and Notices

Signs and notices are common on the real railway, but often absent on a model one. There would normally be a plethora of signs depending on the location. Out on the line there could be speed restriction signs, AWS warning notices, notices for areas of low rail adhesion etc. In a depot there could be signs with instructions for drivers or health and safety notices. Civil engineering such as tunnels, viaducts and bridges often had various signs affixed to them, and there were many signs outside of the railway fence.

Close and careful observation of photographs will reveal a surprising number of signs and notices. There are many commercially available products, but bespoke signs can also be produced on computer, printed at a high resolution and mounted on cardstock. A small piece of microstrip functions as an acceptable post and will suffice for most scenarios.

Industrial Railways

Even as late as the 1970s and 1980s, there were still extensive industrial railways and rail-linked industries in operation in Britain. Industrial railways often operated an eclectic mix of rolling stock. Many of these railways, or at least aspects of them, could make interesting cameos on a main layout, or, in the case of larger industrial sites, be the main focus of a layout in itself.

Another interesting, but seldom modelled industrial railway is the scrapyard. Various rail connected scrapyards existed, such as the large Woodhams Yard at Barry, to smaller outfits that won contracts to scrap perhaps a single locomotive or handful of wagons. Such locations often had their own shunting engine, normally a small industrial diesel type. Empty wagons would be brought in by a mainline locomotive and left to be loaded. The shunting engine would move the wagons around the yard. As there was no 'standard' type of scrap yard, the modeller is free to add one as a cameo to an unused corner of a larger layout, or perhaps develop a scrapyard as a standalone model in its own right.

Industrial railways often featured abandoned machinery, wagons and even locomotives, sometimes left where they had failed or derailed, as it was simply not cost effective

This atmospheric photograph was taken at Attercliffe scrapyard in 1985. A variety of half-scrapped locomotives, gas cylinders (for the cutting torches) plus general detritus can be seen. (*Dave Peachey*)

One thing often not replicated on models of collieries is how dirty and grimy everything looked. Coal dust covered everything, even seemingly the trackside vegetation. This photograph taken at Manvers Colliery in 1986, showcases to good effect the level of general muck and grime. (*Dave Peachey*)

Derailed and abandoned wagons were a common sight on industrial railways and colliery branchlines, as this example at Bamburgh in 1989, shows. (*Dave Peachey*)

Stanton works in Derbyshire operated well into the 1990s. In this photograph, taken in 1986, ladles filled with hot water are being backed into the new works building. Elements of the real railway could be easily incorporated into a model; the modeller does not have to model the entirety of an industrial railway in order to represent one. (*Dave Peachey*)

Even as late as 1989, small steam locomotives could be found in industrial service. Here, CEGB 2, a Robert Hawthorne 7818, gives cab rides on a brake van on an open day at Castle Donnington Power Station. Also, note the extremely clean HAA hoppers in the background. (*Dave Peachey*)

to recover the vehicle or undertake on-site repairs. Again, this is a feature rarely depicted on layouts.

There were also a number of interesting features associated with industrial railways, such as wagon tippers, loading chutes, wagon pulleys and the extensive use of obsolete traction and rolling stock, including small steam locomotives some of which operated well into the 1980s.

Railway Buildings

The structures on a model railway layout are important as they not only add interest for the observer, but they also help to provide an indicator as to the purpose of the railway and the reason that it was originally built.

Realistic and well-constructed buildings do much to provide a sense of time and place on a layout and should be carefully chosen

With the widespread adoption of colour light signalling and power signal boxes, many signal boxes fell into disuse. A box like this one, seen at Gorsey Bank in 1985, would make an interesting model, such dereliction rarely being depicted on layouts. (*Dave Peachey*)

Another item rarely modelled, but very much a feature of the era is the completely derelict but intact station. Shottle Station is on Wirksworth branch in Derbyshire and is seen here in 1987. Closed to passengers in 1949, the intact platform complete station buildings can clearly be seen. The former station yard was used by an oil distribution company (note the storage tanks). The saloon parked at the station is part of a special train for the National Railway Museum. (*Dave Peachey*)

Duncan Hunnisett's 'Shirebook' seen here bathed in natural light illustrates the importance of getting your buildings correct. The buildings, combined with natural sunlight, are very difficult to distinguish from those on the prototype, an incredible achievement when you consider the diminutive size of the model. (*Duncan Hunnisett*)

This photograph shows that drab coloured clothing, browns, blacks and greys, tended to predominate. Glasgow Central, 1984. (*David Ford*)

whilst a small rural station would be more realistically represented with very little ethnic diversity. Fashion obviously mirrored the rest of society at the time, although what is noticeable in photographs taken in the period is the fact many passengers, but particularly women, would travel in their best clothes.

Another aspect to consider is the presence of workers/commuters who fit the time and location being modelled. For example, a large group of businessmen in suits waiting for their first-class business service from London Paddington may look a little out of place on a small suburban station platform set in the Midlands.

Also, you may wish to consider seasons when selecting passengers for your layout; it may look odd if you mix figures wearing thick winter coats, with figures wearing shorts, for example.

Maintenance of Way Staff

Maintaining the railway was an expensive process and on the steam railway such work was very labour intensive. The result was that you would expect to see a lot of people on and around the railway line.

Until the 1970s, practice largely followed the principles of the steam era, being largely unmechanised. This was due primarily to the fact that labour was still relatively cheap and therefore an excess of manpower was never a problem.

For each section of track there was a lineman, who was responsible for inspecting a section of track and effecting minor repairs. He was required to walk his section every day looking for faults such as a loose fishplates, damaged rail chairs, mis-aligned track, subsidence or rock faces in danger of falling. A solitary figure alongside the track would be a reasonable representation of such a worker on a small layout.

The hard manual work of repairing or adjusting a length of faulty track was done by teams platelayers, with a supervisor, or ganger, in charge. To protect the men as they worked a flag man was positioned further up the track, equipped with a flag and a horn. The horn was used as it was a distinctive sound.

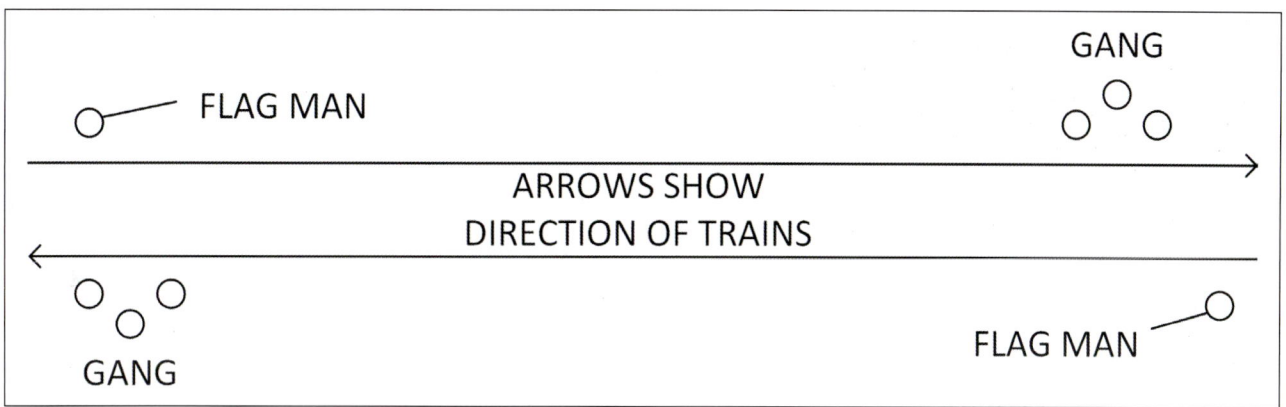

Disposition of typical track gang.

A relaying job at Bolton in 1985, shows to good effect the types of equipment, dress and layout of personnel during such work. Note the flag man at the bottom right of the photograph. (*David Flitcroft*)

Prior to the 1970s, track workers were not issued with standard uniforms and therefore wore their own clothes. Many men wore flat caps, with very few going bareheaded, and almost all wore stout boots to protect their feet from the heavy work. The colours of clothing would reflect the common types of workwear in use at the time, blacks, dark blues, greys and browns tended to predominate. From 1974 onward, improvements to health and safety legislation meant that track workers were increasingly issued with more equipment. In addition, fewer men (particularly those of the younger generation) wore flat caps, although workers could be seen wearing such headwear well into the 1980s.

In the mid-1960s, experiments were conducted with high visibility orange clothing. The first simple orange jerkins were introduced in the early 1970s. By the mid-1970s, the dayglo type plastic waistcoats were standard issue for track side workers, although the mandatory

Track gang hard at Work near Dunton in 1978. Note the complete lack of high visibility clothing. (*David Ford*)

Track welders working at Bottesford in 1983. Note the universal trolley loaded with equipment, a feature not often seen on layouts. (*John Ford*)

wearing of such jackets appears not to have been rigorously enforced.

By the early 1980s, the wearing of hard hats was becoming increasingly common, particularly on jobs involving working in proximity to cranes. The early 1990s saw the introduction of strips of reflective tape onto the vests which soon became ubiquitous throughout the industry and remains so in the modern era.

Track workers would use a variety of tools and equipment, from shovels and picks to welding gear, specially manufactured trolleys and motorised personnel and equipment carriers such as the 'Wickham' type trolley. Some of these items are available commercially, some can be found on 3-D printing websites, and some will best scratch built.

The primary issue with modelling track workers is placing them in a realistic manner; by necessity they have to be working on the track. Trackworkers are best modelled in non-action poses, standing at the lineside (or walking along it) as the trains pass by.

An alternative, if you wanted to model the full extent of track gang in action, you could have them working on a siding or passing loop instead of the main line. This has an additional advantage that the track itself need not even be physically connected to the rest of the layout.

Depot and Yard Staff

Maintenance depots could be busy places with a lot of staff. Typically, those working on locomotives and rolling stock would wear thick overalls or boiler suits. Supervisors and other junior managerial staff sometimes wore a dark blue or brown dust coat to protect their clothes.

Those involved in shunting operations would wear either a high visibility jacket/jerkin, or, in some cases a high visibility type of dust coat. Before the 1980s, the amount of high visibility clothing worn would be a lot less than later years, in a similar manner to trackside workers.

This type of dress would also extend to any other staff seen around yards and sidings, such as fitters, drivers or guards, though of course they would wear a high visibility jacket over their normal working dress which would vary considerably depending on their role.

An outward loading being prepared for dispatch in Bulmers sidings, 1988. Note the shunter wearing a high visibility dust coat. (Jamerail/Flickr)

Chapter 8
Realistic Operations

Movement with a Purpose

It is easy to operate a model railway, we simply place trains on the track and drive them as far and as fast as you would like. However, the real railway doesn't operate like this. Therefore, if we wish to portray our layouts in a realistic manner, we must be cognisant of the real railway and how, where, and why rail vehicles move.

In every scenario on the prototype, movement is undertaken with a specific purpose. It may be something simple such as moving a locomotive in or out of the maintenance shed, or it could be complex, where a train is being assembled by a shunting locomotive for onward dispatch to a specific location. Generally, a minimum number of moves would be undertaken to achieve the objective safely; complex moves were never undertaken if there was a simpler and safer alternative.

Prototypical Movements

A review of layouts at an exhibition will reveal a surprising number, even those modelled to exceptionally high standards; that do not take into account how trains start and stop and how they move on the real railway. It is all too common to see trains starting and stopping in impossibly short distances, In addition, it is common to see trains moving unrealistically, i.e., not considering the real-world rules that all drivers had to follow in the course of their day-to-day duties.

The introduction of Digital Command Control (DCC) has increased the viability of lifelike operation considerably. Detailed discussions about DCC and its capabilities is outside the remit of this volume, but with careful programming, tweaking of the inertia values and the use of onboard sound (particularly the train horn), very convincing operations can be performed.

In the real world, even a small locomotive weighs many tonnes, and cannot start and stop on a sixpence. Similarly, observance of speed limits is a mandatory requirement, such limits being imposed due to track geometry, risk, and the overall condition of the permanent way. Therefore, it makes sense for the modeller to be cognisant of the following aspects when operating their train:

- The type of train, and the maximum permissible speeds of not only the traction unit, but also any coaches and wagons in the train. Unfitted freight trains ran at a maximum speed of 35mph, but sometimes much slower.
- The maximum permitted line speed, including appropriate reductions for the negotiation of turnouts or crossings. Remember that goods lines had a lower maximum speed limit compared to those used for passenger traffic.
- Any features on the route, such as level crossings.
- The distance required to stop a heavily laden train compared to a lighter one.

- The effort required to move a train from a standing start, particularly a lengthy or heavily laden one.

Approximate real to model speeds for n gauge models, rounded to the nearest foot and inch per second

Prototype Speed (MPH)	Prototype (Feet Per Second)	Model Speed (Inch Per Second)
5	7	1
10	15	1
15	22	2
20	29	2
25	37	3
30	44	3
40	59	5
50	73	6
60	88	7
70	103	8
80	117	9
90	132	10
100	147	12
125	183	14

Note that locomotives assigned to the civil engineers were restricted to a maximum speed of 60mph, regardless of their design top speed.

There were standard speed restrictions when a train was operating across common infrastructure features. These were listed in the Sectional Appendix and were typically as follows:

Speed	(mph)
On double lines when passing through junctions between parallel lines or through crossover roads, or when entering or leaving Slow, Good, Loop, Platform or Bay lines.	15
On single lines when passing through loop connections.	20
When receiving, delivering, or exchanging Train Staff or Electric Tokens by hand.	10*
When receiving, delivering, or exchanging Train Staff or Electric Tokens by means of lineside delivery of receiving apparatus.	20*
When receiving, delivering, or exchanging Electric Token by means of automatic exchange apparatus.	25
When travelling overs Goods Lines, Good Loops or Passenger Loops	40
* In the case of Multiple Units or Light Locomotives the train must be stopped	

Finally, trains usually slow down when entering curves, often putting on power as they leave it. As a general rule, the tighter the curve the lower the permitted speed. If you are modelling a real location, the sectional appendices for the era being modelled usually have permitted line speeds. If you are modelling a fictional location, you can draw your own track plan and devise your own speeds but refer to similar real locations for an idea of the types of limits that might be imposed.

Correct Use of The Train Horn

There were certain times when the driver of a train was required to sound their horn. Some of the specific cases are dealt with elsewhere in this chapter, but the following general rules applied:

- When passing a track gang working an open possession. The train sounded the horn, which was acknowledged by the gang lookout. This confirmed to the driver that the lookout was aware that there is a passing train and would warn the track gang accordingly.
- When passing a whistle board (large 'W' in a circle).
- When moving within the confines of a depot, where personnel might be walking across or adjacent to the track. The horn blast was acknowledged by the person(s) walking near the train. If they did not

acknowledge the blast, the driver would repeat the blast until acknowledged.
- When entering and leaving a tunnel.
- Passing through a station with a train on the adjacent platform.
- The driver was at liberty to sound their horn at any other time they felt there was a potential danger.

If your layout and locomotives are equipped with DCC sound, then relatively realistic horn operation can be achieved. Use caution when operating such equipment in an exhibition/public setting as you may cause a nuisance.

Train Behaviour at Level Crossings

Level crossings are another layout feature where many modellers tend to not follow prototype practice. In basic terms, the two types of level crossing that trains were required to negotiate would be the standard type where trains were not required to stop, or the more unusual crossings that incorporated a stop instruction.

The following outline gives the rules for crossings where a train was not required to stop:

- Upon passing the warning board, the driver would regulate their trains speed in order to observe the local speed restriction as detailed on the approach speed restriction board. Upon passing the board, the driver would check that the crossing was clear and that the white light adjacent to the crossing was flashing (indicating that the road traffic signals on the crossing were working properly). The train would then proceed to the crossing at the speed indicated on the restriction board (or lower) and then accelerate as soon as the train reached the crossing. The train would be stopped if any of the following conditions were present:
    a) The white light adjacent to the crossing was not flashing.
    b) The crossing was obstructed.
    c) The crossing could not be seen to be clear due to fog, snow, failure of crossing illumination or other reason.
    d) The movement would compromise Engineers on-track equipment which could not be relied upon to actuate track circuits.
    e) The movement was being made 'wrong direction' and controls for such movements were not provided.

In these scenarios, the driver would stop, ensure it was safe to proceed across the crossing, then move towards the crossing sounding the horn continuously until the train was on the crossing. The outline below gives the rules for instructions where the train was required to stop:

- Upon passing the warning board, the driver would regulate the train speed so that they would be able to stop at the STOP board.
- The plunger (if provided) would be actuated to operate the traffic signals, but only if the train was ready to depart.
- Prior to passing the stop board, the driver would check that the crossing was clear and that the white light adjacent to the crossing was flashing. After sounding the horn, the train would pass over the crossing.
- If the crossing could not be seen to be clear or the white light was not flashing the driver (or another member of the traincrew) would physically check that the train was safe to proceed prior to moving off. The horn would be sounded continuously until the train was on the crossing.

Basic Operations

The use of Passing Loops and Lie-By Sidings

An important manoeuvre on the full-size railway was the passage of two trains travelling in different directions on the same single-track line. This would normally require a passing loop (See Chapter 6). To use a passing loop Train A arrives in Platform 2 and waits for Train B to enter Platform 1 prior to moving off again. Alternatively, because the line is now clear,

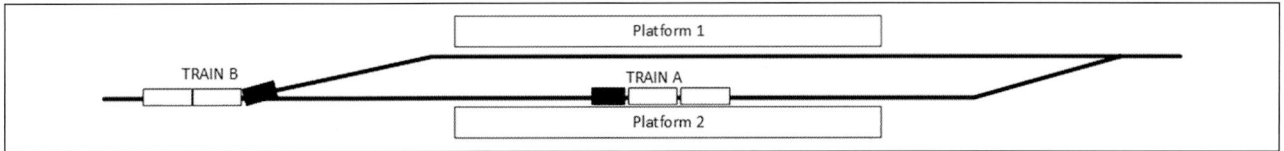

Operation of a passing loop.

Train B can pass through the station without stopping. Although loops were often located at stations, they could also be found at certain locations out on the line.

Another frequent operation was the use of a refuge or a lie-by siding. Such sidings were primarily used to hold slow trains whilst allowing a faster train to pass. Normally, lie-by sidings and refuges were only used to hold freight trains, because passengers would take a dim view of being unceremoniously dumped into a siding to allow another train to pass!

The slow train would be backed into the lie-by siding well ahead of time, allowing the fast train to overtake. Once the fast train had passed, the slow train would be allowed to continue its journey. Of course, all of these moves would be undertaken under the watchful eye of the signalman, and the trains involved would be protected by the system of signalling and interlocking.

Running Round of Passenger Trains at Stations

The most basic of train operations at a station is the running round of a locomotive from one end of the train to the other, to enable the train to depart in the opposite direction. Even with the widespread introduction of multiple units and fixed formation HST sets, there were a preponderance of locomotive hauled trains well into the 1990s.

We start with the arrival of a locomotive hauled passenger train which pulls into the main platform. After a suitable period of time has elapsed to represent the alighting of passengers, the locomotive reverses slightly to push the coaches back into the run around loop. The locomotive then reverses across the crossover onto the loop line. The brakes on the coaches are applied the moment the locomotive is disconnected.

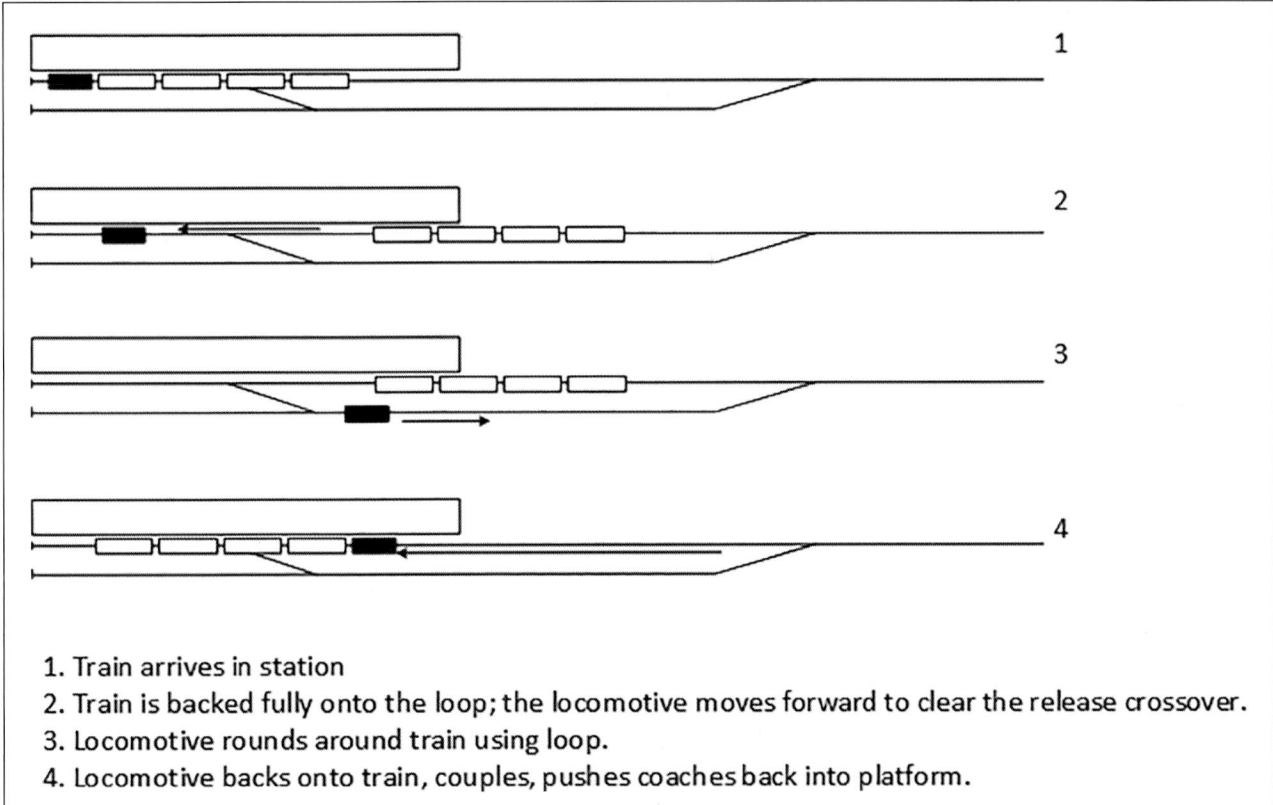

1. Train arrives in station
2. Train is backed fully onto the loop; the locomotive moves forward to clear the release crossover.
3. Locomotive rounds around train using loop.
4. Locomotive backs onto train, couples, pushes coaches back into platform.

Operation of a run around loop.

The locomotive runs out beyond the station limit and pauses whilst groundstaff reverse the points. The loco then reverses back onto the train and the coupling and brake hoses are reconnected. The train is then pushed back into the station and passengers can board.

The bay platform on most stations was used by multiple units, or sometimes for stabling locomotives. It was, however, possible to work a locomotive hauled train in and out of a bay platform, although it required some fairly complex moves.

The train would first have to be shunted into the run around loop to enable the locomotive to run around the train. This would normally be carried out by the station pilot (usually a small diesel shunting locomotive).

In some locations, particularly large main line termini, and even as late as the 1980s where there were intensive locomotive hauled operations, a spare locomotive might be utilised. In its most straightforward form, the locomotive would wait in a stabling spur to await the next arrival. Although originally utilised so that steam locomotives could be coaled, watered and turned, by the 1970s, locomotive services were all hauled by diesel or electric locomotives which did not require turning or extensive servicing, so that method of working was used solely to provide a speedier departure for the train.

Firstly, the train pulls into the bay platform. The spare locomotive is shown waiting at the stabling point. The spare locomotive pulls

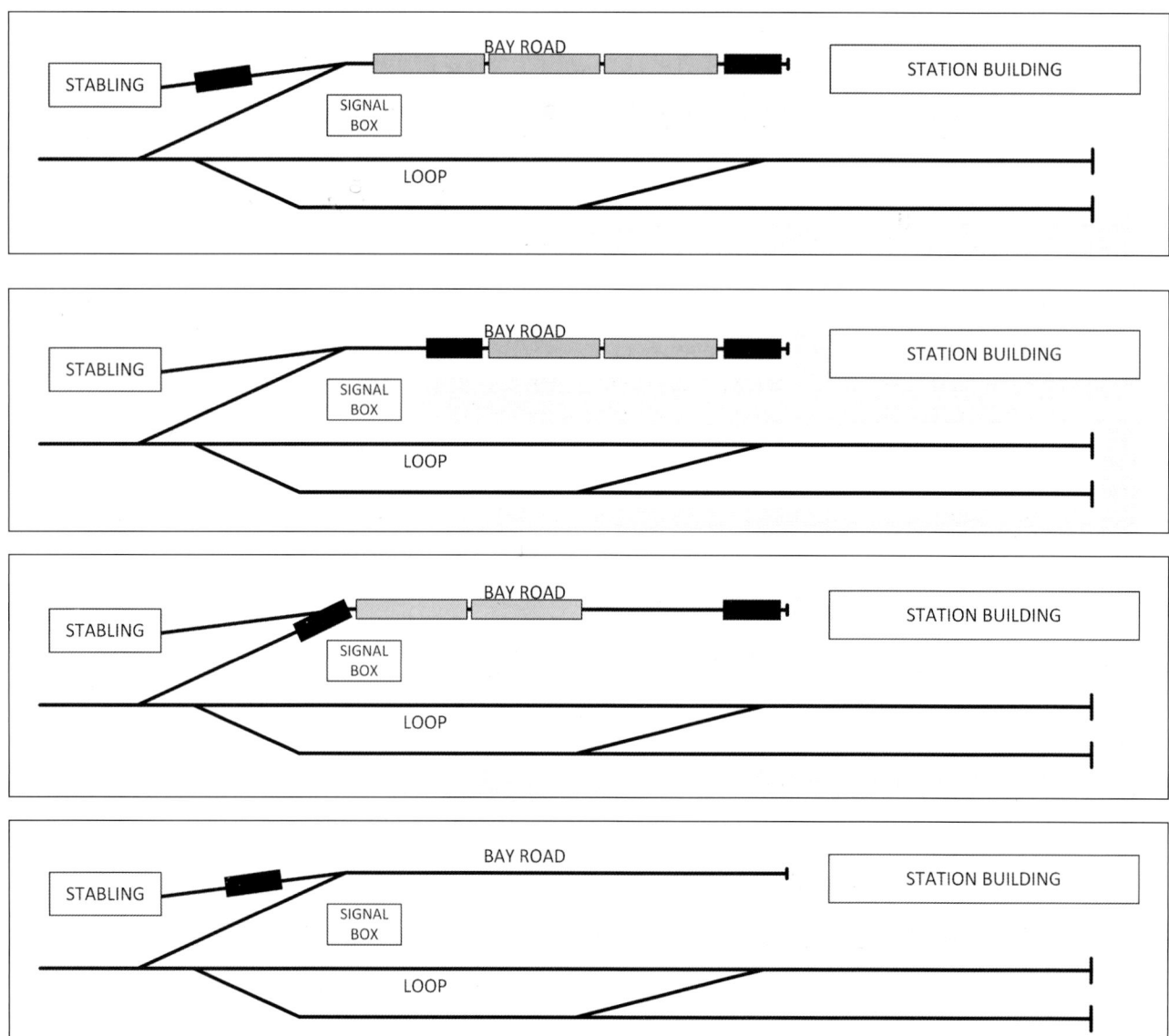

Utilisation of a spare locomotive for speedy departure from terminus.

forward clear of the points and couples to the train. The first locomotive is now uncoupled from the train and the locomotive handbrake is applied.

The train now has right of way and is free to depart. The first locomotive is now spare, free to move onto the stabling point where fuel and boiler water can be topped up and the crew can have their break whilst waiting for the next train and the cycle repeats.

Carriage Sidings

Carriage sidings were used to store the large number of passengers coaches between duty turns, where they could be cleaned, serviced and formed into rakes ready for service. Although intensively used on the steam railway, carriage sidings were in decline by the late 1970s, but could still be found in close proximity to important stations.

Trains would arrive in the station, and the locomotive would be detached, run round (if required) and would trundle off for servicing or stabling. A shunting engine would then collect the rake of coaches and move them to the carriage sidings for servicing and cleaning. If a rake of coaches was required for service, the reverse would take place. The downside of carriage sidings is the space required, even in N, a significant amount of space is required for a convincing layout. An alternative that takes up less space is a carriage shed with the suggestion of extensive carriage sidings behind it.

In some locations, a spare platform or centre road was used to store one or more rakes of coaches. If space is a factor, then this is worth considering. The use of a shunting locomotive to move the rake between platforms within station limits would remain the same.

Shunting

Shunting is a term used to describe a multitude of movements on the railway. It is used to describe the movement of a complete train from one line to another, the positioning of locomotives in a maintenance depot and, most commonly, the movement and sorting of freight wagons at marshalling yards.

In order to understand why shunting was required on the railway we must understand the nature of railway traffic. In the railway world, the word 'traffic' had many meanings. When people travel by train, they were referred to as 'passengers' by the railway, and therefore the management of such trains was termed 'passenger traffic'. Whilst passengers themselves did not require sorting, the coaches in which they travelled sometimes did, with specific coaches often being attached or detached from the train during a long journey. A freight train might need to be shunted several times during its journey, although the amount of freight vehicle shunting in this era was much reduced from the steam age, several large yards were still in operation and the shuffling of wagons to form trains remained common.

There were many types of shunting, flat, loose, hump, fly or rope shunting, although most shunting was carried out by locomotives, most typically small locomotives designed specifically for the duty. In some areas, particularly wagon repair depots, ropes, capstans and 'pinch bars' would be used to move individual wagons. Sometimes specially modified tractors were used, these having large steel plates fitted to avoid damage.

The technique known as 'fly shunting' involved a locomotive pushing wagons ahead of it and then coming to an abrupt stop. The wagons, which would continue to roll, would then be diverted into the required siding. The wagons were not coupled to the locomotive and thus were not fully under control. Consequently, this practice was generally prohibited unless there were no other means of performing the shunting task. Fly shunting is difficult to replicate on an N gauge model due to the light weight of the wagons and the difficulty of coming to an abrupt stop.

Hump shunting involved shunting over an artificial hump. Wagons were pushed to the top of the hump, and then, as they rolled down the other side due to gravity, they would be directed into the appropriate siding. Groundstaff would run alongside the wagons and use brake sticks to apply the handbrake. This was a dangerous practice, so most hump yards used mechanical retarders to slow the wagons down. Hump shunting is another shunting method that is difficult to replicate in N gauge, as the wagons are light, and stopping the wagons in the correct place is difficult, not to mention most hump shunting yards are very large and would require significant space.

The most common type of shunting carried out was known as flat or loose shunting. Marshalling yards were designed to receive incoming trains, break them up and reassemble the wagons into trains for onward travel. Even in the early 1990s, some large yards still existed where such work was undertaken.

Typically, a train would arrive on the arrival or reception line, the locomotive would be detached and be moved to the depot for refuelling and maintenance. The wagons would be moved by a shunting locomotive to the headshunt, shunting line or shunting neck (all three names for the same thing). Normally the shunting locomotive would move the wagons back along the shunting line whilst the head shunter would examine the wagon labels and work out where to divide the train. Once this was completed the first 'cut' would be uncoupled and the driver would then move the locomotive forward. Once the wagons were rolling the driver would apply the brakes. The rest of the train would be stopped, whilst the uncoupled cut would roll forward. Other shunters would have already set the road, and the loose wagons would roll into the correct siding. If required, shunters would run alongside the loose wagons and pin the handbrakes down with a brake stick to slow them down if they were moving too fast.

Occasionally, the shunting team would be required to close up rakes of wagons in sidings which sometimes bunched up near the end of the siding where the shunting was taking place and could potentially foul other roads. This resulted in the need to push the wagons further along the siding in order to provide room for additional wagons. To ensure that the wagons did not roll out of the siding, the first two or three wagons would always have their handbrakes applied to prevent unintended movement.

The realistic stop-start-stop-start of a shunting locomotive engaged in loose shunting is not easy to replicate, but the basic sequence can be followed to provide a representation of the process. It's important that the operator remembers to leave pauses between movements to account for the pinning of handbrakes, setting of roads etc.

A particular problem is the method by which to couple and uncouple wagons. This is an issue in N Gauge due to the size of the models. Coupling wagons together using the standard coupler is straightforward because the couplers are designed to interlink when pushed together. This can be achieved by the operator or on the layout using a locomotive.

More problematic is the issue of uncoupling. There are two basic approaches, one involves the use of track mounted uncoupling apparatus, the other involves the 'hand of God' or representative shunters pole.

The track mounted approach involves a mechanically actuated or magnetic uncoupling ramp. The function of this equipment is to uncouple two vehicles at a predetermined point. The mechanical ramp is operated by a small lever adjacent to the track, the magnetic approach uses a magnet in reverse polarity to small magnets fitted to the bottom of the coupler, when the coupler passes over the magnet, the two opposing poles push the coupler upwards and uncouple the two vehicles. There are also electromagnetic uncouplers from the companies such as Kato which can be energised and deenergised as required. The major issue with this approach is that uncoupling is a fairly

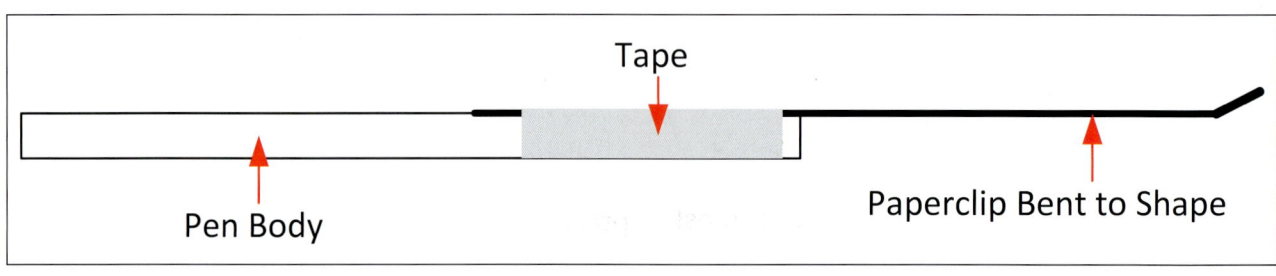

DIY shunter's pole construction.

hit and miss affair, and the fixed nature of the uncoupling apparatus restricts operational flexibility. In addition, the equipment can look rather unrealistic.

An alternative is more in keeping with prototype practice, whereby a 'shunter's pole' is constructed which can be used by the operator to lift up the standard Rapido type coupler and thereby uncouple vehicles. Some imagination and suspension of belief is required, but if you imagine yourself shunting a full-sized wagon and using an appropriate shunters pole, then the illusion is somewhat more palatable. A shunters pole can be as simple or as complex as you would like. The body of a pen, with a paperclip attached, bent into a suitable shape is all that is really required.

Block Working Using Multiple Operators

When two or more people attempt to operate a layout, miscommunication and accidents are bound to happen if a robust communications protocol is not established. As we are attempting to replicate the real railway in miniature, it is sensible to consider replicating the real word practice of block bells used on the traditionally signaled railway system to pass trains between sections controlled by more than one individual.

The simplest way to achieve this is to divide the layout up into sections and assign a sector of the layout to a single operator. Within this sector, this operator has complete control, including the trains, signals and any other control mechanisms. Communication between two operators is achieved via a close replication of prototype block signalling methods.

For a technical description of the principles of block signalling, refer to Chapter 5. There are essentially two separate but linked elements, the block instruments and the block bells. The bell itself is very simple, a basic bell can be used, assuming the operators are within earshot of one another.

Various bell codes are used to signify the conveyance of a certain message. This system of communication was used to ensure that both signalmen understood one another, reducing the chance of miscommunication which could result in an accident.

Bell codes varied over the years and regions, but a common set of bellcodes are listed in the appendices. Many of these codes wouldn't apply to a model railway.

You may even decide to come up with your own list of codes, which, depending on the nature of your railway, wouldn't be essentially incorrect.

Worked Example

We wish to send a single locomotive (light engine) down a section between Signal Box A and Signal Box B (This would typically be represented by two operators on a small to medium sized layout).

The correct sequence would run thusly:

- Box A sends '1' meaning 'Calling Attention'
- Box B sends '1' meaning 'I am Listening'
- Box A sends '2-3' meaning 'I have a light engine coming your way'
- Box B sends '2-3' meaning 'okay, send me the light engine'
- Once the train passes the first signal box (A)
- Box A sends '2' meaning 'Train entering section'
- Box B sends '2' meaning 'okay, the train is on its way to me'

Block working using multiple operators can be as simple or as complex as the layout requires and the operators are capable of undertaking. It goes without saying, that if such working is adopted for an exhibition layout, then extensive practice and training is required to ensure that all operators understand the principles associated with block working, and that mistakes are kept to a minimum.

Developing a Working Timetable

The basic rule is 'if in doubt follow the prototype', the principle being that if we replicate what took place on the real railway in the period modelled, then we will arrive at a somewhat realistic representation.

Table 9: Timetable 1 – Basic timetable Between Stations A and B

	Train A	Train B	Train B	Train A	Train A	Train B
Town A	Dep 06.00		Dep 08.20		Dep 10.40	
Town B	Arr 07.10		Arr 09.30		Arr 11.50	
Town B		Dep 06.00		Dep 08.20		Dep 10.40
Town A		Arr 07.10		Arr 09.30		Arr 11.50

Creation of a timetable for your layout will assist in ordering the passage of trains; it will take on increasing importance as a layout increases in size and complexity. The hardest part of creating such a timetable is knowing where to start. The easiest route is to use timetables from the period, but one may wish to take things a step further and create their own detailed timetable. Such a complex and vast subject could appreciably form the basis of an entire volume and is outside the scope of this book, however, an example is given below for a relatively simplified layout between two towns.

The Use of Wagon Cards

Shunting on the real railway was undertaken purely for operational purposes, the ultimate goal was to earn revenue for the railway.

In comparison, shunting operations carried out on model railways tend to be a haphazard affair, carried out for operator interest, rather than for any specific operational purpose. Shunting in a prototypical fashion, i.e., ensuring that the right wagon is in the right place at the right time, is a challenging activity that will engage those willing to attempt to tackle it.

The mechanics of shunting have been described previously, but the routing of wagons, i.e., what determined where they went and when, is a more advanced subject which we will give an outline treatment here.

The key component of wagon routing was a wagon load label (sometimes called a waybill). This small piece of paper contained details such as the wagon type, its destination, its payload and weights. Although there had been many advances in the railway by the BR Blue Era, even the introduction of TOPS did little to change the wagon load label. In basic respects, it remained unchanged for most of the twentieth century.

A blank label was provided in the BR 30054/1 Working Manual for Rail Staff and all shunters carried a book of these labels for use in their day-to-day duties.

The labels were attached to the wagon via clips, either located on the vehicle solebar or the vehicle sides. Two identical labels were provided, one per side. This label was the information used by railway staff when trains were being split and reformed. Although

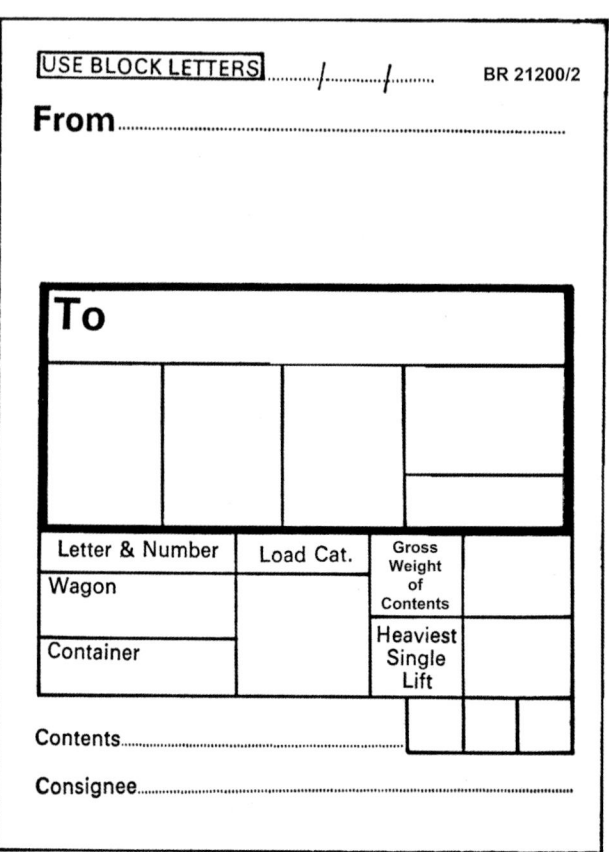

BR 30054/1 wagon load label. (*British Rail*)

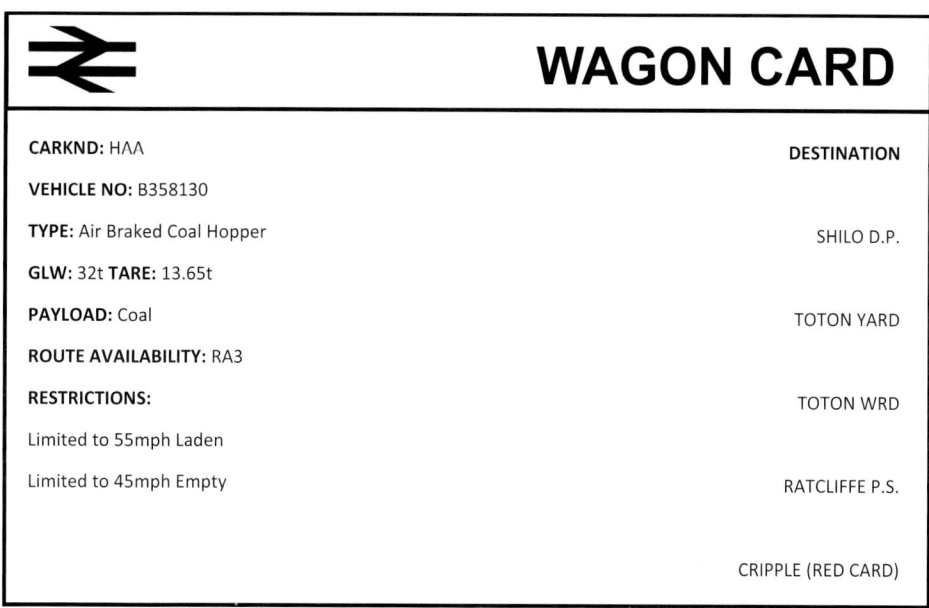

Example wagon card used by the author on a previous layout

in N Gauge we cannot attach these labels to the models, we can still utilise representative cards, either physically printed examples, or electronic examples displayed on a computer or tablet. I prefer the physical label approach as this is more with fitting in the era when computers were largely restricted to specialist work and were not used for general administrative tasks.

To use the system, each model wagon on the layout has their own card. I produced mine on a computer and then printed onto thick paper or thin card. On the card are salient details pertaining to the wagon; type, CARKND, route availability, laden weight (GLW) and so on.

You can customise these details to suit your own requirements. Also on the card is a list of potential destinations. The list of destinations would obviously vary depending on your requirements and your layout. For a simpler or smaller layout, generic destinations would suffice; but for a complex shunting layout, you might want to be very specific, perhaps even down to a specific siding or road. The modeller may also wish to incorporate a 'cripple' status which would simulate a wagon that had been 'red carded' by groundstaff due to a defect. Some modellers may like to also have a 'front' to the card which could repeat some of the details, and perhaps even have a picture of

	Non-Passenger Carrying Coaching Stock (Cont.)
N	Siphon G (News)
O	Covered Carriage Truck (CCT)
P	Covered Carriage Truck (CCT) (BRUTE)
Q	Parcels and Miscellaneous Van
R	Special Parcels Van (SPV) (Ex-Fish Van)
S	Post Office Sorting Van
T	Post Office Storage Van
U	Post Office Brake Van
V	Motor Car Van (2 Tier)
X	General Utility Van (Motorail)
Y	Exhibition Van

	Bogie Steel Carrier
A	Bogie Steel Air Braked 40 ft 77.5 – 78.5 tonne
B	Bogie Steel Air Braked 50 ft 73.5 – 75.5 tonne
C	Bogie Bolster C 30.5 – 42.5 tonne
D	Bogie Bolster D 42 – 57 tonne
E	Bogie Bolster E 32.5 tonne
H	Bogie Bolster H 30.5 tonne
K	Bogie Steel Air Braked 75.5 tonne
O	Bogie Pipe Carrier 30.5 tonne
P	Boplate E 42.5 – 58.5 tonne
Q	Bogie Bolster Q 30.5 tonne
R	Borail 51 tonne
T	Bogie Bolster T 30.5 tonne

	Covered Bulk Carrier
B	Covhop Air Braked 31 tonne
C	Covhop Sand 24/24.5 tonne
G	Grain 20.5 tonne
H	Covhop 24.5 tonne
P	Presflo 17.5, 20.5 and 22.5 tonne
Q	Prestwin 20.5 tonne
S	Ash 21.5 tonne
X	Gunpowder 7/11 tonne
Z	Alumina, Sugar 17.5/23.5 tonne

	Flat
B	Conflat Air Braked 25.5 tonne
E	Conflat E 42.5 tonne (inc. Coke/Clay)
F	Freightliner Inner 52/62 tonne
G	Freightliner Outer (All Types) 52/62 tonne
H	Lowliner 49/63 tonne
I	BR RIV flat (Carfit) 20.5 tonne
J	Freightliner Inner – With Buffers and Drawgear 45 tonne
L	Conflat L 14 tonne
M	Freight Flat 42 tonne (Timber Floor)
P	Conflat F 22.5/31.5 tonne
S	Conflat ISO 20.5 tonne
V	Carflat 10 tonne
W	Bo-flat 42.5 tonne
Z	Other Conflats

	Hopper
A	Coal Hopper Air Braked 32.5 tonne
B	Mineral Hopper Air Braked 32.5/38 tonne (45mph empty)
C	Coke Hopper 20.5 tonne
D	Coal Hopper 33.5 tonne
E	Mineral Hopper Air Braked 32.5 tonne (60mph empty)
J	Ironstone Hopper up to 26 tonne
K	Ironstone Hopper over 26 tonne
S	Coal Hopper S 13 tonne
T	Coal Hopper 21.5/25 tonne
U	Coal Hopper 24.5/25 tonne

	Continental Ferry
C	Tanks – Privately Owned – All Types – 20 – 50 tonne+
P	Flats – Railway Owned inc. Bogie Flats – 15 – 45.4 tonne
I	Interfrigo and Refrigerated Vans – All Types – 13 – 25.6 tonne
L	Large Vans (inc. Motor Car and Large Bogie Vans) 21.5 – 26 tonne

M	Medium Vans (inc. Sliding Roof and Motor Car Vans) 19 – 26 tonne
N	Luggage Vans 7 tonne
O	Open High 27.5/29.5 tonne
P	Privately Owned (except Tanks) 10 – 110 tonne
S	Small Vans 20 – 23 tonne
T	Transfesa 25 – 29 tonne
X	Specials 25 – 60 tonne

Bogie Coil	
A	Bogie Strip Coil 42.5 tonne
E	Bogie Strip Coil E 44.5 tonne (Nylon Hood)
G	Bogie Strip Coil G 44.5 tonne
I	Bogie Strip Coil (RIV) 53.5 tonne with Nylon Hood
K	Bogie Strip Coil K 61 tonne (Nylon Hood)
M	Bogie Strip Coil 42.5 tonne
P	Bogie Strip Coil P 32.5 tonne
R	Bogie Strip Coil R 32.5 tonne
T	Bogie Strip Coil T 61 tonne (Nylon Hood)
U	Bogie Hot Rolled Coil U 32.5 tonne
V	Bogie Strip Coil V 61 tonne (Nylon Hood)
X	Bogie Wide Strip Coil X 40.5 tonne
Y	Bogie Strip Hot Coil 42.5 tonne
Z	Bogie Slab Coil 46 tonne

Two-Axle Coil	
A	Strip Coil A 21.5 tonne (Nylon Hood)
B	Strip Coil B 24.5 tonne (Nylon Hood)
C	Strip Coil C 24.5 tonne
D	Strip Coil D 24.5 tonne
E	Coil E 21 tonne
F	Strip Coil F 24.5 tonne
G	Pig Coil 30.5 tonne
J	Strip Coil J 24.5 tonne
L	Strip Coil L 22.5 tonne
N	Strip Coil 13 tonne
R	Rod Coil R 24.5 tonne
S	Rod Coil 13 tonne
T	Strip Coil T 31.5 tonne
Y	Hybarcoil 12/13 tonne

Mineral	
C	Mineral 16.5 tonne
D	Mineral 21.5/25 tonne
E	Mineral 25 tonne
P	Mineral Ferrous 32.5 tonne
S	Tippler 26.5/27.5 tonne
T	Tippler 24 tonne
X	Mineral 16.5 tonne (Push Type Brakes)

Open	
A	Open Air Braked 31.5 tonne
B	Open Air Braked (Turnover Bolster) 30.5 – 31.5 tonne
C	Open Air Braked (with Bolster) 31.5 tonne
H	High – Steel Sided – inc. Hybar 13 tonne
I	RIV Tube 26.5 tonne
J	High 21.5 tonne
L	Low 13 tonne
U	Shochood B 20.5 tonne
W	High – Wooden Sided – inc. Hybar, Hyhood, Hyrood 13 tonne

Ex-Coaching Stock Departmental Vehicle	
P	Dormitory/Messing/Canteen/Staff
Q	Tool Van
R	Stores & Materials
S	Miscellaneous Operating Vehicle (Runner, Ramp, Pre-Heating, HST Barrier Coach)
T	Service Department Brake Van
V	General Equipment Carrier inc. Breakdown Vehicles
W	Self-Propelled Viaduct Inspection Vehicle, Inspection Saloon
X	Specialist Equipment Vehicles (Instruction/Inspection Coach, Mobile Office, Laboratory, Tunnel Repair, Compressor Van etc.)
Y	Electrification Equipment Vehicles

Privately Owned (Except Tank)

A	Covered Bulk, 2-axle (Covhop, Grain, Lime) 20.5 – 37.5 tonne.
B	Covered Bulk, Bogied (Covhop) 64/71 tonne
C	Cement 2-axle (Presflo) 22-38.5 tonne
D	Presflo Bogied 79 tonne
E	Tip Air 26 tonne
F	Bogie Plate, Weltrols and Carflats 10 – 54 tonne
G	Hopper 2-axle (Aggregate, Gypsum, Ironstone, Salt) 32.5 – 39 tonne
H	Hopper, Bogie (Aggregate, Ironstone, Limestone) 28 – 75 tonne
I	Privately Owned RIV (Carflat, Covhop, Well Wagon) 49.5 – 101 tonne.
J	Cartic 8 tonne
K	Artic 3-axle 9.5 tonne
L	2 Tier Car Carrier 12 tonne
M	Mineral 16.5/22 tonne
N	Open Pallet 40 tonne
O	Open (High) Soda Ash 13 – 37 tonne
P	Weed Killing/Escort Coach
S	Tippler, 2-axle (Ironstone) 20 – 34.5 tonne
T	Tippler, Bogied (Ironstone) 77.5 tonne
V	Van (inc. Palvan) 12 – 36.5 tonne
W	Van Bogied (Palvan) 49/57 tonne
X	Miscellaneous & Special 11 – 96.5 tonne (Bogie Bolster, Bogie Ingot, Match, Ramp, Road-railer, Torpedo, Tube, Single Bolster, Bogie Steel Air Braked)

Railway Operating Vehicle

A	Traffic brakevan (all types) 20 tonne
B	Barrier vehicle
F	Freightliner, Tippler and LTE Adaptors
R	Runner vehicle
T	Brake Tender

Two-Axle Steel Carrier

A	Steel Air Braked 31.5 tonne
B	Steel ABB 31.5 tonne
E	Steel Air Braked 30 tonne
M	Ingot Mould 27.5 tonne
O	Pipe 12 tonne
P	Plate 21.5/31.5 tonne
T	Tube 20.5/22.5 tonne
W	Twin Bolster 31 tonne

Privately Owned Tank

B	Bogie Tank 70 – 79 tonne GLW
C	Bogie Tank 80 – 89 tonne GLW
D	Bogie Tank 90 – 99 tonne GLW
E	Bogie Tank 100+ tonne
I	British RIV Tanks – All Types – 30 – 80 tonne +
M	3-Axle Tank 20 – 39 tonne GLW
R	2-Axle Tank 20 – 39 tonne GLW
S	2-Axle Tank 30 – 39 tonne GLW
T	2-Axle Tank 40 – 49 tonne GLW
U	2-Axle Tank 50+ tonne GLW

Open Bulk Traffic Vehicle

C	China Clay inc. Clay Hood 13 tonne
P	Pig Iron 20.5 tonne
R	Pig Iron 30.5 tonne
S	Sand 13 tonne
T	Timber 17 tonne
U	Timber Pulp 22.5 tonne
Y	Anhydrite 25.5 tonne
Z	Oxide 25.5 tonne

Covered Van

A	Van – Full Length Doors – 20.5 – 30 tonne (Ventilated or Non-Ventilated)
B	Van – Full Length Doors – 28.5 – 30 tonne (Non-Ventilated)
C	Van – Centre Doors – 29 – 30.5 tonne (Non-Ventilated)
D	Van – Centre/End Cupboard Doors (Non-Ventilated) 24.5 – 31 tonne

TOPS CODES FOR ROLLING STOCK • 195

E	Vanwide with Roller Bearings 12 tonne	
F	Fish 12 Tonne	
G	Van – Full Doors (Non-Ventilated) – 25 tonne	
I	BR RIV Ferry Van, Scenery Van, Motor Car Van, 14 – 25 tonne	
M	Vanwide, Plain Bearings, MOD 12 tonne	
P	Palvan 10 tonne	
Q	Palvan 22.5 tonne	
S	Shocvan 12 tonne	
V	Van 11.5/12 tonne	
W	Van with 9 foot Sliding Doors 12 tonne	

Special Purpose Vehicle	
A	Armour Plate (all types) 40.5 – 56 tonne
C	Concrete Beam 32.5 – 51 tonne
D	Flat ED 12 tonne
F	Flat EV, WC or WS 46 – 51 tonne
I	BR RIV Lowmac EU 25 tonne Flatrol EAC 21.3 tonne Rectank 38.5 tonne Weltrol EJC 56.5 tonne
J	Atomic Flask Carrier 77.5 tonne
K	Flatrol (Nuclear Flask) MJ, MJJ 54 – 56 tonne
L	Lowmac (all except RU) 15-25.5 tonne
M	Cartic 32 tonne
N	Flask Carrier 2-Axle 34 tonne
P	Flatrol ED, ELL, ET, MCC, MDD, MHH, WX 25.5 – 81.5 tonne
R	Rectank 38.5 tonne (FD, MD, WC)
S	Concrete Beam Spacer
T	Trestrol (all types) 40.5 tonne
V	Trestle AB or ED 42.5 – 43 tonne
W	Weltrol EL, MC, MV, MX, WBB, WGG, WP, WW 20.5 – 40.5 tonne
X	Twin Girder 20.5 tonne
Y	Boiler, Transformer MA 61 – 81.5 tonne
Z	Flat EU, Transformer MC, Weltrol EN, ENN 100.5 – 135 tonne

Bogie Departmental Vehicle	
A	Bogie Rail/Sleeper – 'Dolphin' 30.5/40.5 tonne
B	Bogie Rail/Sleeper – 'Sturgeon' 51 tonne
C	Bogie Ballast/Sleeper – Drop Sides – 'Pilchard' 20.5 tonne Halibut 52 tonne
D	Skip Storage 46 tonne 'Skate'
G	Bogie General Materials – 'Seacow' 'Sealion' 'Walrus' 40.5/41 tonne
H	Bogie Ballast Hopper, Side & Centre Chutes – 'Whale' 51 tonne
J	Tracklayer
K	Bogie Long Welded Rail – 'Marlin' 'Manta' 14 tonne
L	Bogie Rail – 'Gane A' 35.5/40.5 tonne
M	Bogie Rail – Salmon' 51 tonne
N	Bogie Rail/Bolster ex Traffic – inc. 'Prawn' 'Shrimp' 'Whelk' 30.5 – 57.5 tonne
O	Crane – Bogie
R	Stores & Materials
S	Miscellaneous Operating Vehicles (Runner, Match, Ramp)
T	Service Department, Bogie Brakevan
V	General Equipment Carriers inc. Breakdown Vehicles
X	Specialist Equipment Vehicles (Relaying Equipment, Tunnel Repair, Bridge Repair etc.)
Y	Electrification Equipment Vehicles

Two Axle and Other Departmental Vehicle	
A	Ballast/Sleeper Drop Sides – Medium 13/13.2 tonne
B	Ballast Flat, Side and End Doors – 'Grampus' 'Lamprey' 'Hake' 20 – 24 tonne
C	Ballast/Sleeper Drop Sides – 'Haddock' 'Ling' 'Minnow' 'Plaice' 'Sole' 'Starfish' 'Tunney' 10 – 22.5 tonne
D	General Materials (Other Than Highs and Minerals) 8 – 33.5 tonne
E	Ballast Hopper, Centre Chutes, inc. 'Catfish' 19.5/21.5 tonne
F	Ballast Hopper, Side and Centre Chutes 'Dogfish' 'Trout' up to 25.5 tonne

Two Axle and Other Departmental Vehicle (Cont.)	
G	General Materials 10 – 21.5 tonne
H	General Materials 16.5 – 27.5 tonne
I	Breakdown Crane
J	Ballast Hopper, Side Tipping 'Mermaid' 14 tonne
K	General Materials (DCE) ex 16.5 tonne Minerals
L	Ballast Hopper, Centre Chutes – 'Herring' 20.5 tonne
M	Ballast Hopper, Centre Chutes – 'Mackerel' 17.5 tonne
N	CCE Steel Carriers 13 – 22 tonne
O	Travelling Crane Fixed Axle
P	Dormitory / Messing
Q	Tool Van
R	Stores & Materials
S	Miscellaneous Operating Vehicles
T	Service Department Brake Van
U	Service Department Plough Brakevan – Shark
V	General Equipment Carriers inc. Breakdown Vehicles 'Whiting' & 'Winkle. Weedkiller.
W	Self-Propelled Ballast Cleaner with Buffers & Drawgear
X	Specialist Equipment Vehicles (Ballast Cleaners, Emergency Control Unit, Generator Vehicle, Bridge & Tunnel Repair, Relaying Equipment etc.)
Y	Electrification Equipment Vehicle
Z	Cranes, Snowploughs, Generator (ex Loco)

TOPS Third Letter Codes	
A	Air Braked
V	Vacuum Braked
X	Dual Braked (Vacuum / Air)
B	Air Braked with Vacuum Pipe
W	Vacuum Braked with Air Pipe
O	Unfitted (Handbrake only)

Table 10: TOPS Locomotive Classification Data

Class	Weight (Tonnes)	Brake Force (Tonnes)	ETH Index	RA	Max Speed (MPH)
03	30	13	-	1	28
04	32	14	-	2	27
05	31	13	-	2	18
06	37	15	-	5	23
07	42	21	-	6	20
08	49	19	-	5	15
09	49	19	-	5	27
13	120	38		8	20
17	69	35	-	4	60
20/0	73	35	-	5	75
20/9	35	73	-	5	60
24/0	80	38		6	75
24/1	74	38		5	
25/0	74	38		5	90
25/1	74	38		5	90
26/0	78	35		5	80

Class	Weight (Tonnes)	Brake Force (Tonnes)	ETH Index	RA	Max Speed (MPH)
31/4	107	48	66	6	90
33	76	35		6	85
37/0	105	50		5	90
40	133	51		6	90
44	133	63		7	90
45/0	136	63		7	90
45/1	136	63		7	90
46	139	63		7	90
47/0	117	60	-	6	95
47/3	117	60	-	6	95
47/4	117	60		6	95
47/6	117	60		6	95
50	117	59		6	100
55	100	51		5	100
56	128	60		7	80
58	129	60		7	80
59	126	69		7	60
60	129	74	-	7	60
71	77	41		6	90
73/0	75	31		6	80
73/1	75	31		6	90
74	85	41		7	90
76	89	43		8	65
81	80	40		6	100
82	77	38		6	100
83	76	38		6	100
84	77	38		6	100
85	80	41		6	100
86/0	80	40		6	100
86/1	80	40		6	100
86/2	80	40		6	100
87/0	80	40		6	100
87/1	80	40		6	100
90/0	84.5	40	95	7	110
90/1			-		75
91	81.5		95	7	125
92	126		180	8	87

Appendix B
Freight Stock Markings

Marking	Significance
Diagonal White Stripe on Side of Mineral Wagon	Position of end door
Two short white lines in the form of a 'V' at bottom centre of wagon side	Bottom doors
Vertical white stripes, 3 on each side and ends	Equipped with shock-absorbing gear
Large solid yellow triangle pointing upwards on side of 25t mineral wagon	To distinguish from 21.5t mineral wagon
Axle boxes painted yellow, with or without red stripes	Fitted with roller bearings
White star/stars on underframe	Position of vacuum brake release cord
Solid white or black triangle pointing downwards on solebar or wagon side	Fitted with two vacuum brake cylinders and manual changeover gear to adjust brake for empty or loaded running
Solid white or black triangle pointing downwards on solebar or wagon side	As above, but unnecessary to pull release cord before operating manual changeover gear
Letters 'S' or 'SS' on Continental Ferry Wagons	Denotes suitability for running and certain speeds on Continental railways.
The sign 'CL' on the side of a bogie bolster wagon	Indicates position of centre line
Letters 'RIV' enclosed within a rectangle on side of vehicle	Conforms to requirements for running over Continental railways
Anchor surrounded by a rectangle on side of Continental ferry wagon	Conforms to the loading gauge agreed by the International Union of Railways
Top half of circle, with a cross at either end, on side of wagon used for international traffic	Must not be allowed to pass over a shunting hump
Red triangle on side of WR china-clay wagon	Interior lined with zinc
Letter 'L' on side of WR china-clay wagon	Longitudinal floorboards
Length measurement between arrows on side of Continental ferry wagon	Indicates wheelbase or distance between bogie pivots

Appendix C

Loads Permitted with Specific Brake Forces

This table applies to Class 4 and Class 6 trains and is taken from the 1977 reissue of BR 30054/6 'Working Manual for Rail Staff'.

Available Brake Force (Tonnes)	*Maximum Speed (mph)*						
	45	50	55	60	65	70	75
	Load (Tonnes)						
30	140	130	115	100	90	80	75
40	190	170	150	135	120	110	100
50	240	210	190	170	150	135	125
60	285	255	230	200	180	165	150
70	330	295	270	235	215	190	175
80	380	340	305	270	245	220	200
90	430	385	345	305	275	245	225
100	475	425	385	340	305	275	250
110	525	470	420	370	335	300	275
120	570	510	460	405	365	330	300
130	620	550	500	440	395	355	325
140	670	595	535	475	425	385	350
150	715	640	575	510	455	410	375
160	765	680	615	540	485	435	400
170	810	720	650	575	515	465	425
180	860	765	690	610	545	495	450
190	905	805	730	645	575	520	475
200	955	850	765	680	605	550	500
210	1000	890	805	710	635	575	525

Available Brake Force (Tonnes)	Maximum Speed (mph)						
	45	50	55	60	65	70	75
	Load (Tonnes)						
220	1045	935	840	745	665	600	550
230	1095	975	880	780	695	630	575
240	1140	1020	920	810	725	655	600
250	1190	1060	955	845	755	685	625
260	1235	1105	995	880	790	710	650
270	1285	1145	1035	915	820	740	675
280	1335	1190	1070	945	850	765	700
290	1380	1230	1110	980	880	795	725
300	1430	1275	1150	1015	910	820	750
310	1475	1315	1185	1050	940	850	775
320	1525	1360	1225	1080	970	875	800
330	1575	1400	1265	1115	1000	905	825
340	1620	1445	1300	1150	1030	930	850
350	1670	1490	1345	1185	1060	955	875
360	1720	1530	1380	1220	1090	985	900
370	1765	1575	1420	1250	1120	1015	925
380	1810	1615	1460	1285	1150	1040	950
390	1860	1660	1500	1320	1180	1070	975
400	1905	1700	1540	1355	1210	1095	1000
450*	2145	1910	1735	1525	1360	1230	1125
500*	2385	2125	1925	1695	1510	1370	1250
550*	2620	2335	2120	1865	1660	1510	1375
600*	2860	2545	2315	2035	1810	1645	1500
650*	3100	2760	2505	2205	1960	1780	1625
700*	3340	2980	2700	2370	2110	1920	1750
750*	3580	3190	2900	2545	2260	2060	1875
800*	3815	3400	3090	2715	2410	2200	2000
850*	4050	3620	3280	2880	2560	2335	2110
900*	4290	3830	3480	3050	2710	2475	2230

* Loads over 2000 tonnes apply to air braked trains only.

Appendix D
Locomotive Life by Class

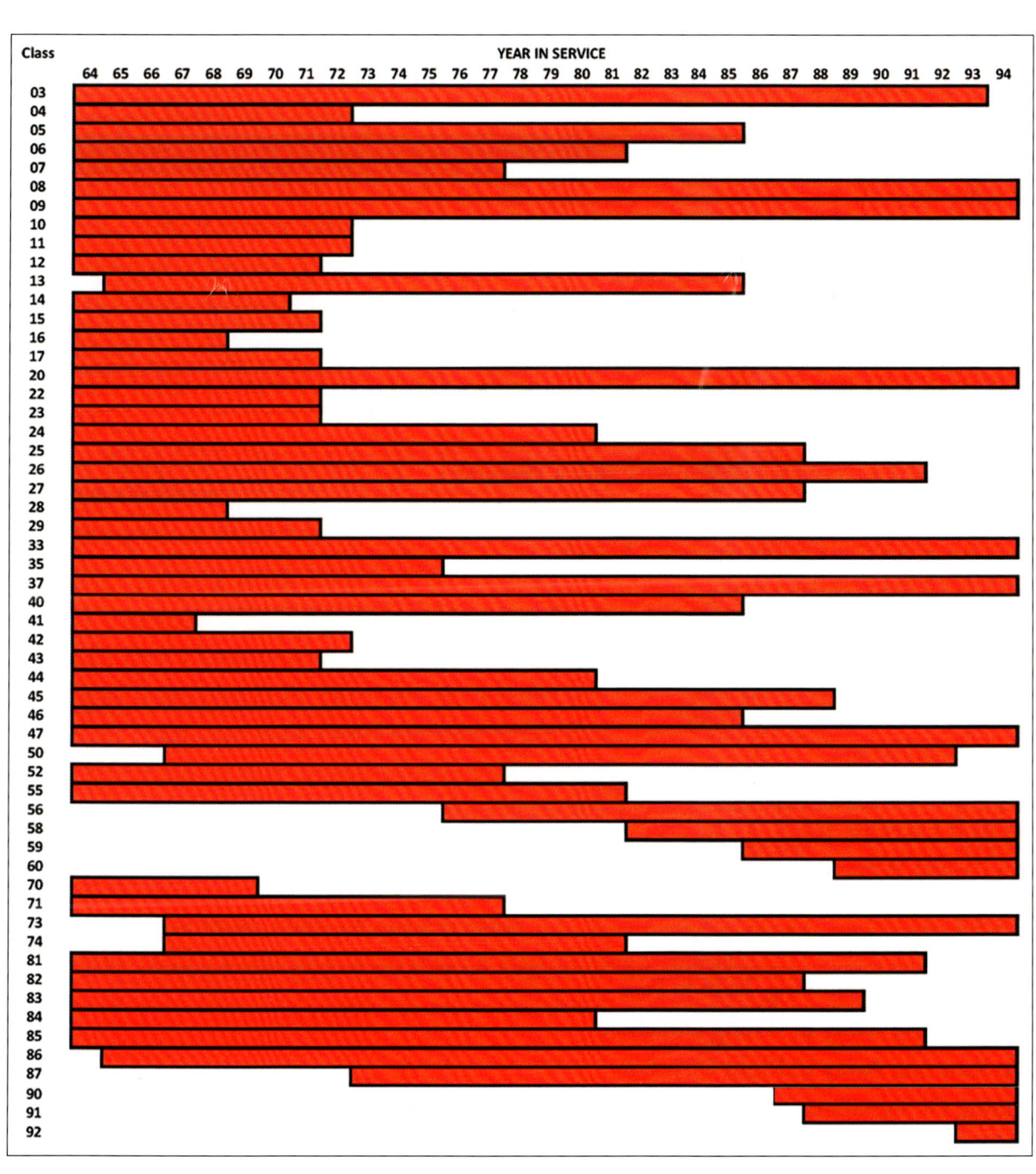